MACMILLAN SECO

Physics
○○○○○○○○○○○○○○○○○○○○

Student's Book 4

David Glover

Consultants:
Sam Kisala
Hezron Mwangi

MACMILLAN
KENYA

HIRANI

Macmillan Kenya (Publishers) Ltd
Kijabe Street, PO Box 30797, Nairobi
A division of Macmillan Publishers Limited
Companies and representatives throughout the world

www.macmillan-africa.com

ISBN 9966-945-56-3

Text © David Glover 2006
Design and illustration © Macmillan Publishers Limited 2006

First published 2006

All rights reserved; no part of this publication may be reproduced, stored in a retrieval system, transmitted in any form or by any means, electronic, mechanical, photocopying, recording, or otherwise, without the prior written permission of the publishers.

Designed by Ascenders Partnership
Typeset by A1 Graphics
Illustrated by Thomas Rudman and Tek Art
Cover design by Gary Fielder, AC Design
Cover photographs by Getty Images/The Imagebank (l) front and (l) back and Science Photo Library © David Parker (r) front and (r) back

The authors and publishers would like to thank the following for permission to reproduce their photographs:
Alamy pp 40 © Terry Fincher Photo International, 73
©Elmtree Images
Alan Thomas pp 16(a&b), 75
Corbis p 68(inset) © Lester Lefkowitz
DK Images p 126
Dr Mike Taylor pp 91, 120
MML Electrics p 46(r) © www.leisure-electrics.co.uk
Photographers Direct pp 71(t), 81 © Peter Gould Photography, 71(b) © John Rattle/Sound Snaps, 90 © James Robertson Photography
Science Photo Library pp 14 © Volker Steger, 15 © David Sharf, 46(l) © CNES, 1995 Distribution Spot Image, 66 © Alex Bartel, 68(main) © Garry Watson, 100 © Alan Sirulnikoff, 111 © Patrick Blackett, 118 © Oulette & Theroux, Publicphoto Diffusion
Topfoto p 119
www.hartrao.ac.za p 5 © Michael Gaylard/Hartebeesthoek Radio Astronomy Observatory

Printed and bound in Nairobi by English Press Limited, P.O. Box 30127 - 00100 - GPO, Nairobi - Kenya.

	2009		2008		2007		2006		2005
10	9	8	7	6	5	4	3	2	1

Contents

Preface ... vi

1 Thin lenses

1.1 Types of lenses ... 1
1.2 Ray diagrams ... 2
1.3 Image formation ... 3
1.4 Determination of focal length ... 9
1.5 The human eye and its defects ... 12
1.6 Optical devices ... 14
1.7 Using the lens formula and the magnification formula ... 16
1.8 Telescopes ... 18
Review questions ... 21

2 Uniform circular motion

2.1 The radian, angular displacement, angular velocity ... 22
2.2 Centripetal force ... 24
2.3 Examples and applications of uniform circular motion ... 29
2.4 Solving problems on circular motion ... 32
Review questions ... 34

3 Floating and sinking

3.1 Archimedes' principle and the law of flotation ... 35
3.2 Relative density ... 38
3.3 Applications of Archimedes' principle and relative density ... 39
3.4 Solving problems on Archimedes' principle ... 40
3.5 Project to construct a hydrometer ... 41
Review questions ... 43

4 Electromagnetic spectrum

4.1 Electromagnetic spectrum ... 44
4.2 The properties of electromagnetic waves ... 45
4.3 The detection of electromagnetic radiations ... 45
4.4 Applications of electromagnetic radiations ... 47
4.5 Solving problems involving $c = f\lambda$ ... 49
Review questions ... 50

5 Electromagnetic induction

5.1	Investigating electromagnetic induction	51
5.2	Induced emf (Faraday's and Lenz's laws)	52
5.3	Mutual induction	56
5.4	Generators	57
5.5	Transformers	59
5.6	Applications of electromagnetic induction	61
5.7	Solving problems on transformers	63
5.8	Project to construct a simple transformer	63
	Review questions	65

6 Mains electricity

6.1	Sources of mains electricity	66
6.2	Power transmission	67
6.3	The domestic wiring system	69
6.4	The consumption and cost of electrical energy	75
6.5	Problems on mains electricity	75
	Review questions	77

7 Cathode rays and the cathode ray tube

7.1	The production of cathode rays	78
7.2	The properties of cathode rays	79
7.3	Cathode ray oscilloscope and television tubes	80
7.4	Uses of the cathode ray oscilloscope	82
7.5	Solving oscilloscope problems	85
	Review questions	87

8 X-rays

8.1	The production of X-rays	88
8.2	Energy changes in an X-ray tube	89
8.3	The properties of X-rays	90
8.4	The dangers of X-rays	90
8.5	Uses of X-rays	90
8.6	Solving problems on X-rays	91
	Review questions	92

9 The photoelectric effect

9.1	The photoelectric effect	93
9.2	Factors affecting photoelectric emission	96

9.3	Energy of photons	97
9.4	The Einstein equation	98
9.5	Applications of the photoelectric effect	98
9.6	Solving problems on the photoelectric effect	100
9.7	Project to construct a burglar alarm	101
	Review questions	103

10 Radioactivity

10.1	Radioactive decay	104
10.2	Half-life	106
10.3	Types of radiations and their properties	108
10.4	Detectors of radiation	110
10.5	Nuclear fission and nuclear fusion	112
10.6	Nuclear equations	116
10.7	The hazards of radioactivity	117
10.8	The applications of radioactivity	118
10.9	Solving problems on radioactivity	120
	Review questions	122

11 Electronics

11.1	Conductors, semiconductors, insulators	123
11.2	Intrinsic and extrinsic semiconductors	125
11.3	The p–n junction diode	126
11.4	Applications of diodes: rectification	128
11.5	Project to construct a simple radio receiver	128
	Review questions	132

Appendices

Appendix 1 Safety in the laboratory	133
Appendix 2 Collecting and displaying experimental data	133
Appendix 3 Answering examination questions	135
Appendix 4 SI units	137
Appendix 5 Significant figures	138

Index

139

Preface

This book is the fourth in a series of Macmillan Secondary Physics texts, prepared specifically to meet the needs of the new revised Kenyan Secondary School Curriculum. Each book is written by specialist science educators, teachers and examiners. *Physics Student's Book 4* fully covers all the topics required in the Form Four Physics syllabus and equips the learner with the knowledge and skills required for the KCSE examinations.

The specific objectives of each topic are stated at the beginning of each chapter so that the learner knows what should be achieved by studying that chapter. Concepts and skills are systematically developed to achieve the objectives. Illustrations are used throughout to help with the understanding of concepts and principles. Exercises and questions, including end-of-chapter review questions, give opportunities for pupils to assess their understanding and to practise answering examination questions. *Appendix 3* gives guidance on answering examination questions. Each chapter ends with a summary to reinforce learning.

The book provides a varied set of experiments and activities to help the understanding of concepts and principles and to develop the required skills. Through the investigative approach, the learner will develop the manipulative and observational skills needed for practical examinations. The experiments have been selected carefully and presented through clear, simple procedures. They will help the learner to make accurate observations and draw logical conclusions from the experiments. Where necessary, 'Cautions' have been given to allow the learner to appreciate the need for safety procedures when performing experiments. Labelled drawings, showing experimental set-ups, will help the user to set up equipment safely and to obtain accurate results.

This book will be useful to all Form Four students studying Physics in Kenyan secondary schools.

Note to the teacher on safety

When practical instructions have been given, we have attempted to indicate hazardous substances and operations in the 'Cautions'. Teachers should, however, be familiar with any advice on safety from employers and professional organisations and should follow these requirements at all times. Students should know what to do in an emergency, such as a fire.

1 Thin lenses

Learning objectives

After completing the work in this chapter you will be able to:

1. describe converging lenses and diverging lenses
2. describe the principal focus, the optical centre and the focal length of a thin lens using ray diagrams
3. determine the focal length of a converging lens experimentally
4. locate images formed by thin lenses using the ray construction method
5. describe the characteristics of images formed by thin lenses
6. explain image formation in the human eye
7. describe the defects of vision in the human eye and how they can be corrected
8. describe the use of lenses in various optical devices
9. solve numerical problems involving the lens formula and the magnification formula.

1.1 Types of lenses

Magnifying glasses, microscopes, telescopes, cameras, slide and movie projectors all contain lenses. A lens is a transparent piece of glass or plastic that is shaped to use refraction to converge or diverge the light passing through it. Different shaped lenses refract light in different ways. The most common types of lens have either one spherical and one plane surface, or two spherical surfaces as shown in Fig. 1.1.

Converging (convex) lenses are thicker at the centre than at the edge. Their action is to converge a parallel beam of light to a focus, after passing through the lens.

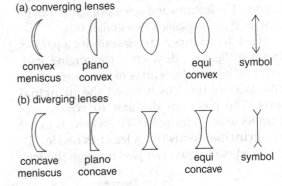

Fig. 1.1 Spherical lenses

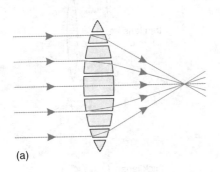

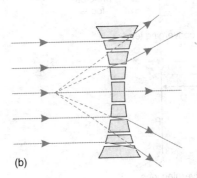

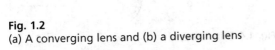

Fig. 1.2
(a) A converging lens and (b) a diverging lens

Chapter 1: Thin lenses

Diverging (concave) lenses are thicker at the edge than at the centre. Their action is to diverge a parallel beam of light.

The converging and diverging actions of lenses can be understood by picturing a lens as a series of prisms that deviate incident light rays. This is shown in Fig. 1.2. The prism at the centre of the lens has parallel faces and therefore does not deviate the ray passing through it. The prisms become more steeply angled as their distance from the centre of the lens increases. Rays incident close to the edge of the lens are thus deviated more than those incident near its centre. In this way, parallel rays are refracted so that they converge to, or appear to diverge from, a single point.

1.2 Ray diagrams

In Form Two you studied the properties of curved mirrors, drawing ray diagrams to explain their actions. The features and actions of lenses can be described and explained in a similar way.

Fig. 1.3(a) shows a ray diagram for a parallel beam of light incident on a converging lens. The line joining the centres of curvature of the two faces of the lens is called the **principal axis**. The point on the axis to which rays parallel to the principal axis converge, is called the **principal focus** (F). A lens has two foci (F and F') because rays can pass through the lens from either side. The distance between the principal focus and the centre of the lens (the optical centre) is called the **focal length**, f.

The corresponding diagram for a diverging lens is shown in Fig. 1.3(b). A diverging lens has two **virtual foci** (F and F'), from which rays incident parallel to the principal axis appear to diverge after passing through the lens.

Spherical aberration

As with mirrors, only rays which are near to the principal axis of a converging lens are brought to a sharp focus. If the curvature of a lens is large compared to its diameter, then **spherical aberration** is a problem; rays relatively far from the principal axis are converged too much and the image is blurred. Such lenses are called wide-aperture lenses. Wide-aperture simple lenses do not produce a sharp image, so lenses of small aperture are usually used. If an optical device has to have a wide-aperture lens then spherical aberration is corrected by using several lenses in combination. The shape and separation of the lenses is such that the aberration caused by one lens is compensated for by opposite aberrations caused by the other lenses. Such lens systems are commonly used in cameras, telescopes and microscopes.

In the theory that follows in this chapter, only **thin lenses** are considered. These are lenses with a curvature small compared to their diameter, for which spherical aberration is negligible.

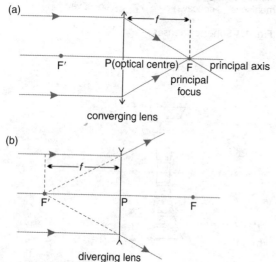

Fig. 1.3 Ray diagrams
(a) Converging lens
(b) Diverging lens

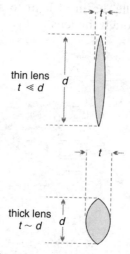

Fig. 1.4 Thin lenses

1.3 Image formation

Activity 1.1 Investigating image formation by thin lenses

You will need:
- a converging lens
- a diverging lens
- a lens holder
- a long pin and a short pin with large heads, mounted in corks
- a needle mounted in a cork.

Method

Investigate the behaviour of the converging lens first, and then the behaviour of the diverging lens.

1. Place the lens 10 cm in front of the needle.
2. Adjust the height of the needle so that an image of its eye can be seen through the lens.
3. View the image of the needle with the aid of one eye only, looking along the principal axis of the lens.
4. Use the pin to find the position of the image of the needle by the **no parallax** method, using either a long pin or a short pin according to whether the image is behind or in front of the lens. Figs. 1.5(a) and (b) show what is seen when the eye is moved to the left when a long pin is (a) behind and (b) in front of the image in a converging lens. When the pin is in the same position as the image, the needle (viewed through the lens) and pin head (viewed over the lens) will appear to move together, staying in line.
5. Increase the distance between the needle and the lens in steps of 5 cm, recording your observations as shown in Tables 1.1 and 1.2 to record: (i) whether the image is upright or inverted; (ii) whether it is magnified or diminished; (iii) whether it is in front of or behind the lens; (iv) whether it is in front of or behind the object; (v) the distance of the image from the lens.

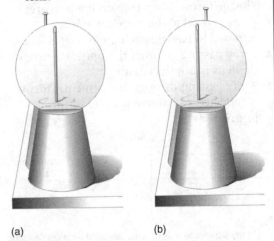

(a) (b)

Fig. 1.5 Investigating image formation by lenses
(a) Long pin behind the image position
(b) Long pin in front of the image position

Table 1.1 Images formed by a converging lens

Distance of object from lens	(i) Is the image upright or inverted?	(ii) Is the image magnified or diminished?	(iii) Is the image in front or behind the lens?	(iv) Is the image in front or behind the object	(v) Distance of image from lens
10 cm	Erect	Magnified	Behind	Behind	10 cm

(continued)

Activity 1.1 (continued)

Table 1.2 Images formed by a diverging lens

Distance of object from lens	(i) Is the image upright or inverted?	(ii) Is the image magnified or diminished?	(iii) Is the image in front or behind the lens?	(iv) Is the image in front or behind the object	(v) Distance of image from lens

Ray construction

The nature and location of the image formed by a given lens can be found by geometrical construction, see Fig 1.6. The rules for locating the image formed by a converging lens are:

1. A ray parallel to the principal axis is refracted to pass through the principal focus.
2. A ray passing through the optical centre of the lens is not deviated.
3. A ray passing through the principal focus is refracted to emerge parallel to the principal axis.

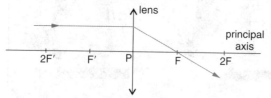

(1) ray parallel to axis of lens, deviated through its focus

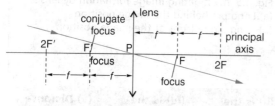

(2) ray through centre of lens, undeviated

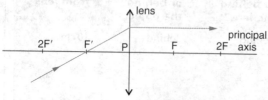

(3) ray through conjugate focus of lens, deviated parallel to its axis

Fig. 1.6 Diagrams illustrating the rules for image formation by a converging lens

To locate an image, only two rays from the object are needed. (But remember that objects actually emit rays in all directions, not just the two chosen for image location.) Use Rules 1 and 2 to obtain the most accurate results. Information about the image, whether it is real or virtual, upright or inverted, magnified or diminished, can be deduced from the ray diagram.

Converging lens

Consider a converging lens with optical centre P and principal foci F and F'. In Fig. 1.7 the object, O, is represented by an upright arrow, while the image is marked I. The thin lens is represented by a straight line down the centre of the lens at which all refraction is assumed to take place.

First, we find the position of the image when the object is more than twice the focal length, f, from the lens. Fig. 1.7(a) shows two rays from the top of the image. The first is labelled (i) because it obeys Rule 1 above. The second is labelled (ii) and it obeys Rule 2. The corresponding point in the image is located where the two rays cross after refraction.

- In (a) the object is placed beyond $2f$. The image is real, inverted, smaller than the object, and is between f and $2f$.
- In (b) the object is moved forwards until it is at $2f$. The image is real, inverted, the same size as the object, and at $2f$ on the opposite side of the lens.
- In (c) the object is placed between $2f$ and f. The image is real inverted, magnified and located beyond $2f$.
- In (d) the object is placed at f. The image is formed at infinity – in practice no image can be seen.

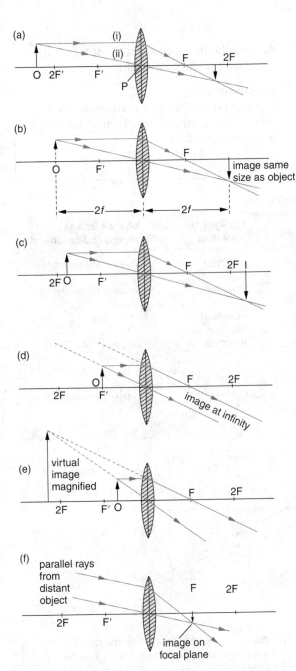

- In (e) the object is placed between f and the lens. A magnified virtual image is formed on the same side of the lens as the object. The image is located by extending the diverging rays backwards to locate the point from which they appear to come.
- In (f) the object is placed at infinity. Rays from a point on a very distant object are parallel. They are converged to form an image of the object in the focal plane of the lens. The image is inverted.

Fig. 1.8 An image of the Sun formed by a converging lens – the Sun is very distant and therefore the image is formed in the focal plane of the lens

The properties of the images formed by a converging lens are summarised in Table 1.3. Notice that real images are always inverted and virtual images are always upright.

Fig. 1.7 Finding the position of an image formed by a converging lens

Activity 1.2 Locating images formed by a convex lens by ray construction

You will need:
- graph paper
- a sharp pencil
- a ruler

Method
Follow the rules outlined above to locate the images formed by a convex lens with $f = 5$ cm for the following object distances:

(continued)

Chapter 1: Thin lenses 5

Activity 1.2 (continued)

$u = 12$ cm, $u = 10$ cm, $u = 8$ cm, $u = 5$ cm, $u = 3$ cm.

1. Draw the principal axis, mark the position of the lens and its focal points F and F'.
2. Draw an object 2 cm high at the required distance from the lens.
3. Construct rays to locate the image (use Fig. 1.7 as a guide).
4. Describe the position and nature of the image for each value of u.

Discussion

It is not possible to locate an image by construction when $u = f$ (5 cm). The image is formed at infinity.

Table 1.3 The properties of images formed by a converging lens

Position of object	Position of image	Real or virtual	Upright or inverted	Size of image compared to object
At infinity	At f	Real	Inverted	Smaller (diminished)
Beyond $2f$	Between $2f$ and f	Real	Inverted	Smaller (diminished)
At $2f$	At $2f$	Real	Inverted	Same size
Between $2f$ and f	Beyond $2f$	Real	Inverted	Larger (magnified)
At f	At infinity	Not visible		
Between f and the lens	On the same side of the lens as the object	Virtual	Upright	Larger (magnified)

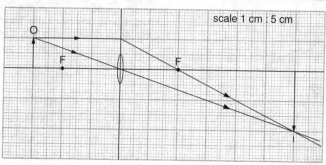

Fig. 1.9

Example

An object of height 5 cm is placed at a distance of 15 cm from a converging lens of focal length 10 cm. Find the position and nature of the image by graphical construction.

Answer

The image is real, inverted and 30 cm from the lens (see Fig 1.9).

Exercise 1.1

Use graph paper to draw a ray diagram to scale for each of the following cases. Draw two rays from each object to locate the image. Describe each image as being real or virtual; inverted or upright; diminished, magnified or the same size as the object.

1. An object of height 1 cm is placed 8 cm in front of a converging lens of focal length 3 cm.
2. An object of height 1 cm is placed 4 cm in front of a converging lens of focal length 3 cm.
3. An object of height 0.5 cm is placed 2 cm in front of a converging lens of focal length 3 cm.

Diverging lens

The ray diagrams in Fig. 1.10 show how to locate an image formed by a diverging lens. A ray from the top of the object travelling parallel to the principal axis is refracted so that it travels on a line which, when projected backwards, passes through the focus F. A ray from the same point on the object passing through the centre of the lens is not deviated. The two rays diverge after refraction by the lens. They are projected backwards to find the point at which they appear to cross. This is the virtual image of the point on the object from which the two rays originated.

The images formed by diverging lenses are always virtual, upright and diminished. They are always on the same side of the lens as the object. Table 1.4 summarises the properties of images formed by diverging lenses, where f is the focal length of the lens.

Example

A burning candle of height 5 cm is placed 10 cm in front of a diverging lens of focal length 15 cm. Use the graphical method to find the position and height of the image.

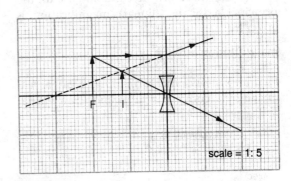

Fig. 1.11

Answer

The image is 6 cm in front of the lens, virtual, upright and 3 cm high.

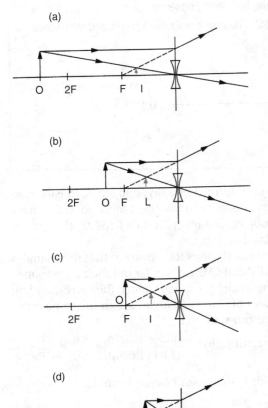

Fig. 1.10 Images formed by a diverging lens

Table 1.4 Properties of images formed by a diverging lens

Position of object	Position of image	Real or virtual	Upright or inverted	Size of image compared to object
At infinity	At f, on the same side of the lens as the object	Virtual	Upright	Smaller (diminished)
Between infinity and the lens	Between f and the lens on the same side of the lens as the object	Virtual	Upright	Smaller (diminished)

Exercise 1.2

1. An object 1 cm high is placed 2 cm in front of a diverging lens of focal length 4 cm. Find, by graphical construction, the position and nature of the image.

2. An object 1 cm high is placed 8 cm in front of a diverging lens of focal length 4 cm. Find, by graphical construction, the position and nature of the image.

3. An object is placed in front of a diverging lens as shown in Fig. 1.12.

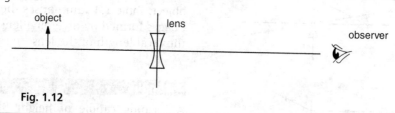

Fig. 1.12

Which of the following statements is correct?

(a) The image is real, inverted and in front of the lens.

(b) The image is real, diminished and upright.

(c) The image is virtual and on the same side of the lens as the observer.

(d) The image is virtual, upright and diminished.

Magnification

For objects that are at right angles to the principal axis, the (linear) magnification is given by:

$$\text{magnification} = \frac{\text{image height}}{\text{object height}}$$

Fig. 1.13 shows the formation of (a) a real image by a converging lens and (b) a virtual image formed by a diverging lens. In each case the distance from the object to the mirror (object distance) is labelled u; the image distance is v.

In both cases it is apparent that the triangles ABP and CDP are similar (all their corresponding angles are equal). Since the corresponding sides of similar triangles are in the same ratio, we can state that:

$$\text{magnification} = \frac{\text{image height}}{\text{object height}} = \frac{AB}{CD} = \frac{AP}{CP} = \frac{v}{u}$$

This is the magnification formula:

$$m = \frac{v}{u}$$

The formula applies to the formation of both real and virtual images by converging and diverging lenses.

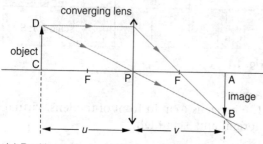

(a) Real image formed by a converging lens

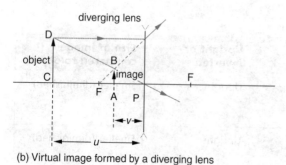

(b) Virtual image formed by a diverging lens

Fig. 1.13 Magnification

Example

An object 2 cm high is 40 cm from a converging lens. An image of the object is formed 20 cm from the lens. What is the magnification and the height of the image?

Answer

Object height = 2 cm; u = 40 cm; v = 20 cm. m = ?; image height = ?

$$\text{magnification} = \frac{\text{image height}}{\text{object height}} = \frac{v}{u} = \frac{20}{40} = 0.5$$

image height = 0.5 × 2 cm = 1 cm

Exercise 1.3

1. An object 2 cm high is 10 cm from a converging lens. The image is 15 cm behind the lens. Find the magnification and size of the image.
2. An object 1 cm high is 2 cm in front of a lens. A virtual image is formed 4 cm in front of the lens.
   (a) Find the magnification and size of the image.
   (b) Draw a ray diagram to scale for this situation.

1.4 Determination of focal length

The focal length of a lens can be determined by three different methods.

Method 1: Estimation

The image of a distant object (one effectively at infinity) is formed in the focal plane of a converging lens. The focal length of the lens can thus be estimated by holding the lens in front of a white screen and adjusting the lens–screen distance until a sharp image of a distant object is formed. The distance between the lens and the screen is then approximately equal to the focal length of the lens.

Method 2: Using the lens formula

In Section 1.7 it will be shown that the object distance, u, image distance, v, and focal length of a lens are related by the lens formula:

$$\frac{1}{f} = \frac{1}{u} + \frac{1}{v}$$

Rearranging this:

$$f = \frac{uv}{u + v}$$

With this formula, f can be calculated from measurements of corresponding values of u and v.

Activity 1.3 Experiment to estimate the focal length of a converging lens

You will need:
- a converging lens
- a piece of card.

Method
1. Stand near a brightly lit window.
2. Hold the lens and card as shown in Fig. 1.14.
3. Adjust the distance between the lens and the card to form an image of a distant object such as the tree shown.
4. Measure the distance between the card and the lens to estimate the focal length.

Fig. 1.14 Estimating the focal length of a converging lens

Activity 1.4 Experiment to determine the focal length of a converging lens using the lens formula

You will need:
- a converging lens in a holder
- an illuminated object
- a white screen
- a ruler.

Method
1. Use Method 1 to obtain an initial estimate of the focal length of the lens.
2. Arrange the object, lens and screen as shown in Fig. 1.15.
3. Set the object distance between f and $2f$ (where f is your estimate from step 1).
4. Adjust the image distance v until a sharp image is formed on the screen.
5. Measure and record u and v as shown in Table 1.5.

(continued)

Activity 1.4 (continued)

6. Repeat steps 2–5 for several values of u, some values less than $2f$ and some values greater than $2f$.

Calculation
- Calculate f for each pair of values of u and v.
- Calculate a mean value for f from all your readings.

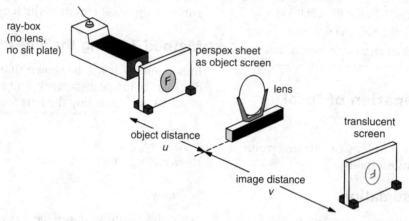

Fig. 1.15 Experimental arrangement

Table 1.5 Results for Method 2

u (cm)	v (cm)	$f = \dfrac{uv}{u+v}$ (cm)

Discussion questions
1. Look at the range of values of f obtained in your table. Discuss what this tells you about the accuracy of your final result.
2. Why could this method not be used with a diverging lens?

In the experiment in Activity 1.4 a real image is formed on the screen. A diverging lens always forms a virtual image which cannot be projected onto a screen. In Activity 1.5 the lens formula method is used to find the focal length of a diverging lens by using the method of no parallax to locate the image position.

Activity 1.5 Experiment to determine the focal length of a diverging lens using the lens formula

You will need:
- a diverging lens
- a lens holder
- a pin with a large head mounted in a cork
- a needle mounted in a cork.

Method
1. Place the lens 30 cm in front of the needle.
2. Adjust the height of the needle so that an image of its eye can be seen through the lens.
3. Use the pin to find the position of the image of the needle by the no parallax method (Fig. 1.16). The image is on the same side of the lens as the object and between the object and the lens. Place the pin so that its head can been seen over the top of the lens. Move the eye left and right and observe the relative movement of the image of the needle eye (seen through the lens) and the pin head (seen over the lens).

Activity 1.5 (continued)

Move the pin towards or away from the lens until there is no relative movement (no parallax) and the pin and needle appear to move together. The pin is now located at the position of the image.

Fig. 1.16 Experimental arrangement

4. Measure and record the object distance u and the image distance v as shown in Table 1.6.
5. Repeat steps 2–4 for several values of u.

Table 1.6 Results

u (cm)	v (cm)	$f = \dfrac{uv}{u+v}$ (cm)

Calculation
- Calculate f for each pair of values of u and v.
- Calculate a mean value for f from all your readings.

Discussion question
Why must the needle eye be viewed through the lens and the pin head over the lens?

Method 3: The lens–mirror method

Fig. 1.17 shows a luminous point object placed in the focal plane of a converging lens. Light from the object passes through the lens and emerges as a parallel beam. The beam is then reflected from a plane mirror back through the lens. The lens converges the beam to form an image of the object. The image is also in the focal plane.

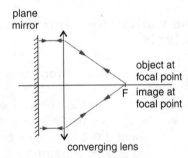

Fig. 1.17 A lens–mirror combination

Activity 1.6 Experiment to measure the focal length of a converging lens using the mirror method

You will need:
- a luminous object (alternatively a mounted pin)
- a white screen
- a converging lens on a stand
- a plane mirror on a stand
- a ruler.

Method
1. Arrange the source, lens and mirror as shown in Fig. 1.18.

2. Adjust the distance between the source and the lens until a sharp image is formed on the screen adjacent to the source.
3. Measure the distance between the source and the lens
4. Repeat the procedure several times, resetting the apparatus on each occasion.

(**Alternative method**: if an illuminated source is not available then replace the

(continued)

Activity 1.6 (continued)

source by a pin. Observe the image of the pin as shown in Fig. 1.18(b). Adjust the distance between the pin and the lens, using the no parallax method to judge when the pin and its image are in the same plane. Record the distance between the pin and the lens. Repeat the procedure several times.)

Calculation
Find the mean value of the focal length.

Discussion
The uncertainty in the final value resulting from random errors is reduced by taking several readings and calculating their mean. The apparatus should be reset each time to ensure that the measurements are independent.

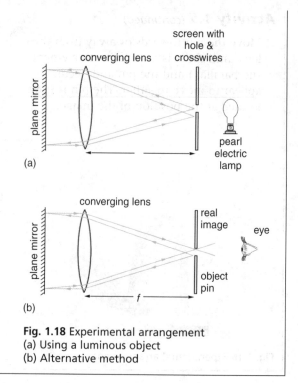

Fig. 1.18 Experimental arrangement
(a) Using a luminous object
(b) Alternative method

1.5 The human eye and its defects

Examine your eyes in a mirror. Draw what you see.

The human eye is similar to a camera in many ways. The eye is roughly spherical with a diameter of about 2.0–2.5 cm. The front of the eye is covered with a transparent layer called the **cornea**. Behind the cornea are the **aqueous humour**, the **iris** and the **lens**.

The aqueous humour is a watery liquid. The iris acts as a diaphragm to control the amount of light entering the eye. The circular opening in the iris is the **pupil**. The pupil appears dark because it reflects little light. The diameter of the pupil varies from about 2 mm in daylight to about 6 mm in darkness. Most of the middle part of the eye is made up of a clear jelly-like fluid called the **vitreous humour**.

Light rays from objects enter the eye and are converged by the cornea and the lens onto the **retina**. The retina contains light-sensitive cells and nerve fibres. The nerve fibres transmit messages to the brain through the **optic nerve**. Although the image formed on the retina is inverted, the brain can interpret it correctly. The region where the optic nerve leaves the eye is insensitive to light. It is called the **blind spot**. When an image falls on the blind spot, the eye cannot detect it.

Most of the bending of light rays is done by the curved cornea and lens. The lens can alter its curvature and hence its focal length. This is carried out by the **ciliary muscles** and **suspensory ligaments**. When a person is looking at distant objects, the lens is at its thinnest because the radial ciliary muscles have contracted, while the circular ciliary muscles have relaxed and the suspensory

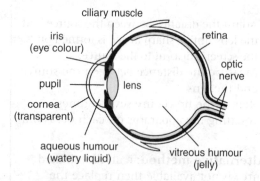

Fig. 1.19 Structure of the human eye

ligament is taut. When the person is looking at near objects, the radial ciliary muscles are relaxed while the circular ciliary muscles are contracted. The suspensory ligament is slack, thus making the lens thicker. This adjustment is called **accommodation**.

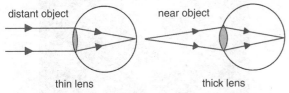

Fig. 1.20 Focusing on distant and near objects

The furthest distance at which an eye can see objects clearly is called the **far point**. People with normal eyesight can focus on objects at infinity. The point closest to the eye at which objects can be clearly seen is called the **near point**. For a normal eye, the near point is about 25 cm from the eye. The near point varies from person to person and with age.

An image persists on the retina for about one-tenth of a second after an object is lost from view. This persistence of vision explains why continuous movement is seen in motion pictures and on television. In a movie film 24 still pictures are flashed onto the screen, one after another, each second and this gives the impression of continuity of movement.

Each eye produces a slightly different image and the brain combines the two into one. This **binocular vision** helps us to judge distance.

Eye defects

Short sight (myopia) is a common eye defect. A short-sighted person can see near objects clearly but not distant objects. The person's far point is closer than infinity. This is because either the focal length of the lens is too short or the eyeball is too long. This means that rays from a distant object are focused in front of the retina. This defect can be corrected by wearing spectacles (or contact lenses) with diverging lenses. The lens diverges the light before it enters the eye so that the image is focused on the retina.

Long sight (hypermetropia) is another eye defect. A long-sighted person can see distant objects clearly but cannot focus on near ones. This is because either the focal length of the lens is too long or the eyeball is too short. The person's near point is further away than 25 cm. This means that light rays from a close object are focused behind the retina. This defect can be corrected by wearing spectacles (or contact lenses) with converging lenses.

Activity 1.7 Investigating spectacle lenses

1. Talk to staff and students who wear spectacles – do they have difficulty in focusing on distant objects (short sight) or on nearby objects (long sight)?
2. Examine their spectacles – are the lenses converging or diverging?
   A quick test to check if a lens is converging or diverging can be made by holding the lens a few millimetres above a sheet of white paper. A diverging lens forms a shadow that is darker than the surroundings (because it diverges the light passing through it); the shadow of a converging lens is brighter than the surrounding paper.

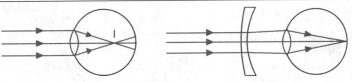

Fig. 1.21 Short sight is corrected by a diverging lens

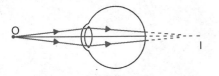

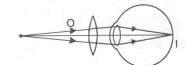

Fig. 1.22 Long sight is corrected by a converging lens

1.6 Optical devices

Telescopes, microscopes and cameras are optical devices that use lenses to produce magnified images of small objects, make distant objects appear closer or record an image by focusing it onto a photographic film.

The magnifying glass (the simple microscope)

A magnifying glass, or simple microscope, is a converging lens of short focal length. The glass is placed at a distance shorter than its focal length from the object. An upright, virtual and enlarged image of the object is formed.

In effect, using a magnifying glass enables the viewer to bring the object closer to the eye than the normal near point of 25 cm. The magnification of a magnifying glass is approximately $\frac{25}{f}$, where f is the focal length of the lens and 25 is the near point of the typical viewer in centimetres.

Fig. 1.23 A magnifying glass in use

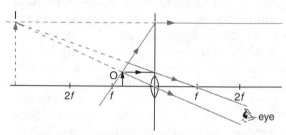

Fig. 1.24 How a magnifying glass forms an image

The compound microscope

A laboratory microscope is sometimes called a *compound* microscope because two short-focus converging lenses are used to magnify an object.

> **Exercise 1.4**
>
> 1. What is the magnifying power of a converging lens of focal length 10 cm for a person with a near point of 25 cm?
> 2. Which makes a better magnifying glass – a converging lens of short focal length or one of long focal length? Explain your answer.

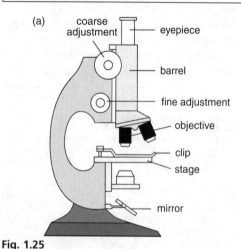

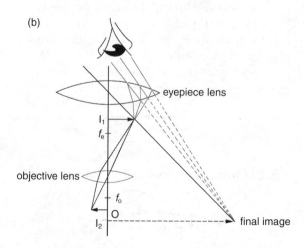

Fig. 1.25
(a) Parts of a microscope
(b) How an image is formed in a compound microscope

The lens closer to the object is called the objective, while the other lens forms the eyepiece.

The object to be viewed is placed on the stage of the microscope. It is illuminated from below by a reflecting mirror or a light source. The object is focused by moving the whole microscope barrel towards or away from the object. This can be done by using the fine and coarse adjustment wheels. Different magnifications can be obtained by changing the objective or the eyepiece, or both.

Fig. 1.25(b) shows how the object is placed just beyond the focal point of the objective lens. The objective produces an enlarged, real image of the object between the two lenses. The eyepiece then acts as a magnifying glass to produce the final enlarged image below the objective. In normal use the microscope is adjusted to give an image at the near point of the viewer's eye.

Exercise 1.5

1. Describe the final image formed by a microscope as compared with the object. List some uses of optical microscopes.

2. Which of the following pairs of lenses are most suitable for a compound microscope?

(a) Two diverging lenses of focal length 100 cm and 5 cm

(b) Two converging lenses of focal length 100 cm and 5 cm

(c) Two diverging lenses of focal length 5 cm and 3 cm

(d) Two converging lenses of focal length 5 cm and 3 cm

(e) Two converging lenses both of focal length 3 cm.

The camera

A camera is a light-tight box with at least one converging lens at the front. The purpose of the lens is to produce a real image of an object on the **film** at the back of the camera. The film contains light-sensitive chemicals and, after exposure to light, it is developed into a negative. From the negative a positive photograph is printed.

A simple camera is focused by moving the lens towards or away from the film. For distant objects, the images are formed on the focal plane. As the object moves nearer to the lens, the image moves further away from the focal plane. Thus, to get the image on the film

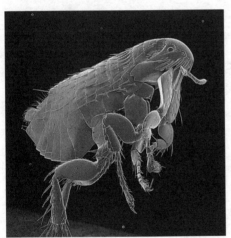

Fig. 1.26 Photomicrograph of flea

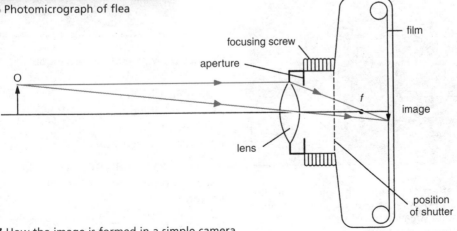

Fig. 1.27 How the image is formed in a simple camera

Chapter 1: Thin lenses

the lens has to be moved further out. The lens is usually moved by turning the **focusing ring** which has a distance scale marked on it.

The amount of light entering the camera can be controlled by the length of time that the **shutter** is open. This is often referred to as exposure time. For fast-moving, or brightly lit objects, the exposure time is very short. For still or dull objects the exposure can be lengthened. Exposure times are usually fractions of a second. A typical exposure time is 1/125 s.

The amount of light entering a camera can also be controlled by opening and closing the **aperture**. The aperture is an opening of variable size just behind the lens. If the photograph is taken in dim light, a large aperture is used. If the light is bright, a smaller aperture is used. The aperture is expressed in terms of *f*-numbers. The *f*-number gives the diameter of the aperture as a fraction of the focal length of the lens. For example, an *f*-number of 8 corresponds to an aperture whose diameter is one-eighth of the focal length of the lens. Hence, the smaller the *f*-number the larger is the aperture. Typical *f*-numbers are 22, 16, 12, 8, 5.6, 4, 2.8 and 2.

There is a range of distances within which objects are captured in focus. This range is called the **depth of field**. The depth of field is controlled by the aperture. A large depth of field means that both near and far objects are in focus at the same time. To obtain a large depth of field, a small aperture is used – for example *f*–22. A large aperture gives a small depth of field and the subject stands out against a blurred background.

1.7 Using the lens formula and the magnification formula

Fig. 1.29 shows the formation of an image by a converging lens.

(a)

(b)

Fig. 1.28 The effect of changing the aperture on the depth of field. Photos taken with
(a) large aperture (small depth of field)
(b) small aperture (large depth of field)

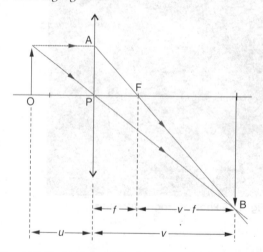

Fig. 1.29 Derivation of the lens formula

PA = object height
IB = image height

$$\frac{IB}{PA} = \frac{\text{image height}}{\text{object height}} = m = \frac{v}{u}$$

Triangles PAF and IBF are similar since all corresponding angles are equal.
Therefore:

$$\frac{IB}{PA} = \frac{IF}{PF} = \frac{(v-f)}{f} = \frac{v}{u}$$

Exercise 1.6

1. An object is placed 15 cm from a converging lens of focal length 10 cm. Find the position and magnification of the image.

2. Find the position of an object if it forms an image 20 cm behind a converging lens of focal length 15 cm.

3. An object is placed 10 cm from a lens. The image is real and twice the size of the object. Find the focal length of the lens and say what kind of lens it is.

4. A luminous object is placed 10 cm in front of a diverging lens. The height of the object is 5 cm. The focal length of the lens is 15 cm. Find the position and height of the image. Is the image real or virtual; upright or inverted?

5. (a) Draw a ray diagram to show how a converging lens works as a magnifying glass.

   (b) An object placed in front of a converging lens produces an image at a distance of 15 cm from the lens and on the same side as the object. Determine the focal length of the lens.

   (c) Fig. 1.30 shows a certain eye defect. Name the defect and draw on a similar diagram an arrangement to correct the defect.

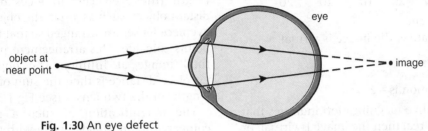

Fig. 1.30 An eye defect

So: $(v - f)u = fv$
$$vu - fu = fv$$
$$vu = fu + fv$$
$$vu = f(u + v)$$
$$f = \frac{uv}{(u + v)}$$
$$\frac{1}{f} = \frac{(u + v)}{uv} = \frac{1}{u} + \frac{1}{v}$$

This is the lens formula:
$$\frac{1}{f} = \frac{1}{u} + \frac{1}{v}$$

We have only derived the formula for a converging lens forming a real image. But, provided the 'real is positive' sign convention (which is explained below) is used, the formula applies to the formation of real and virtual images by converging and diverging lenses.

The real-is-positive sign convention
1. All distances are measured from the optical centre of the lens.
2. All distances to real objects and images are positive; all distances to virtual objects and images are negative.
3. The focal length of a converging lens is positive; that of a diverging lens is negative.

Example 1

An object is placed 30 cm from a converging lens of focal length 20 cm. Find the position and magnification of the image.

Answer
$f = +20$ cm and $u = +30$ cm
$$\frac{1}{f} = \frac{1}{u} + \frac{1}{v}$$
So, $\frac{1}{v} = \frac{1}{f} - \frac{1}{u} = \frac{1}{20} - \frac{1}{30} = \frac{3-2}{60} = \frac{1}{60}$

Thus $v = +60$ cm
Since v is positive, the image is real.
$$m = \frac{1}{u} = \frac{60}{30} = 2$$

So the magnification, $m = 2$.

Example 2

An object is placed 10 cm from a converging lens of focal length 20 cm. Find the position and magnification of the image.

Answer
$f = +20$ cm and $u = +10$ cm
$$\frac{1}{f} = \frac{1}{u} + \frac{1}{v}$$

Chapter 1: Thin lenses

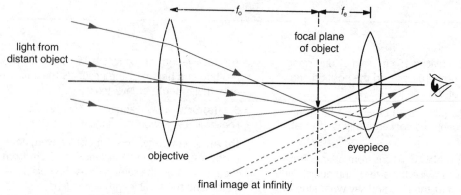

Fig. 1.31 Formation of an image using a refracting (astronomical) telescope

So, $\dfrac{1}{v} = \dfrac{1}{f} - \dfrac{1}{u} = \dfrac{1}{20} - \dfrac{1}{10} = \dfrac{1-2}{20} = \dfrac{-1}{20}$

Thus $v = -20$ cm

Since v is negative, the image is virtual.

$m = \dfrac{v}{u} = \dfrac{-20}{10} = -2$

The magnification is -2.

Note: A negative magnification indicates that if the object is real then the image is virtual, or vice versa. In this case, we are told that the object is real, therefore the image must be virtual. It is generally acceptable to quote only the magnitude of the magnification, i.e. 2 in this case. In this type of question it is a good idea to draw a ray diagram of the situation to check your answer.

 **Telescopes**

Refracting telescopes

A refracting telescope consists of a converging lens with a long focal length (the objective) and a converging lens with a short focal length (the eyepiece). To focus on a very distant object, such as a star, the objective and eyepiece lenses are arranged so that their focal points coincide. This arrangement produces a final image at infinity. The objective to eyepiece distance is then the sum of the focal lengths of the two lenses (see Fig 1.31).

The magnification of such a telescope, compared to an unaided view with the naked eye, is given by:

$$\text{magnification} = \dfrac{\text{focal length of objective}}{\text{focal length of eyepiece}}$$

This formula shows why this type of telescope needs an objective with a long focal length and an eyepiece with a short focal length. The objective forms a real image (I) of a distant object at the principal focus of the eyepiece. The eyepiece treats this real image as an object and magnifies it, forming a virtual and inverted image.

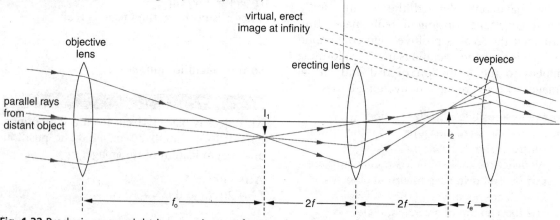

Fig. 1.32 Producing an upright image using a refracting (terrestrial) telescope

Refractors with large objectives (of over 100 mm) are very expensive. Refractors such as these are called **astronomical telescopes** and are good for viewing the moon and planets. Cheaper types, with only one objective lens, may suffer from chromatic aberration, which is the failure of the lens to bring light of different wavelengths (colours) to a common focus. This results in a faint coloured halo around objects. It is corrected by using specially coated lenses and/or a double-objective lens.

Refractors are also good for terrestrial viewing but the upside-down image has to be inverted so that distant objects can be viewed correctly. The inverted image can be made upright with an additional lens as shown in Fig. 1.32 – this type of refractor is called a **terrestrial telescope**.

Reflecting telescopes

The **Newtonian telescope** uses a converging parabolic primary mirror to collect and focus incoming light onto a flat secondary (diagonal) mirror which, in turn, reflects the image out of an opening at the side of the main tube and into an eyepiece. The eyepiece contains one or more lenses of short focal length. The focal length of the primary mirror is about 1000 mm for portable instruments.

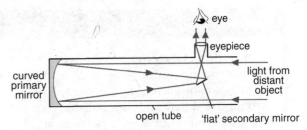

Fig. 1.33 The Newtonian telescope

Such telescopes are excellent for observing deep night sky objects such as remote galaxies, star clusters and nebula. They have only small optical aberrations and deliver bright images. Newtonian telescopes do not suffer from chromatic aberration like refractors. The open tube, however, allows image degrading air currents to affect performance. There is also a slight loss of light due to secondary mirror obstruction when compared with refractors. Newtonian telescopes have the lowest cost per objective diameter of all the telescopes since mirrors can be produced at lower cost than lenses for medium to large apertures. The magnification for reflectors is calculated in the same way as for refractors.

Activity 1.8 *Project to construct a telescope*

You will need:
- lenses
- mirrors
- card/plastic tubes
- adhesive tape
- simple tools.

1. Start by examining the lenses and/or mirrors you have available. Measure their focal lengths and diameters. Depending on the items you have, you may choose either to make a refracting telescope or a reflecting telescope.

2. Plan and sketch your design. Use Figs. 1.31 and 1.33 to help. You should consider:
   - the correct separation of the lenses/mirrors;
   - how to mount the lenses and mirrors in tubes;
   - how to adjust the separation of the objective and eyepiece for focusing;
   - how to mount your telescope tube for use.

3. Construct your telescope and test its performance.

4. Write a project report.

Summary

- The action of a converging (convex) lens is to converge a parallel beam of light.
- The action of a diverging (concave) lens is to diverge a parallel beam of light.
- The curvature of a thin lens is small compared to its diameter; spherical aberration is negligible.
- The following rules are used to locate the position and nature of an image formed by a lens:
  - a ray parallel to the principal axis is refracted to pass through the principal focus;
  - a ray passing through the principal focus is refracted to emerge parallel to the principal axis;
  - a ray passing through the optical centre of the lens is not deviated.
- The image formed by a converging lens may be real and inverted, or virtual and upright. A real image may be magnified or diminished depending on the position of the object; a virtual image is always magnified.
- The image formed by a diverging lens is always virtual, upright and diminished.
- Magnification = $\dfrac{\text{image height}}{\text{object height}} = \dfrac{v}{u}$
- $\dfrac{1}{f} = \dfrac{1}{u} + \dfrac{1}{v}$
- $f = \dfrac{uv}{u+v}$
- The point closest to the eye, at which objects can be clearly seen, is called the near point; for a normal eye, the near point is about 25 cm from the eye.
- Short sight (myopia) is corrected with diverging spectacle lenses.
- Long sight (hypermetropia) is corrected with converging spectacle lenses.
- The magnification of a magnifying glass is approximately $25/f$, where f is the focal length of the lens in centimetres.
- A compound microscope has a short focal length objective lens which forms a real image; an eyepiece lens is used to view and magnify the image.
- A camera lens produces a real image of an object on the film at the back of the camera; the lens aperture affects the amount of light reaching the film and the depth of field.
- The magnification of a telescope (refracting or reflecting) is given by:

  magnification = $\dfrac{\text{focal length of objective}}{\text{focal length of eyepiece}}$.

Review questions

1. (a) Draw a diagram of a thin converging lens to show what is meant by:
   (i) the principal focus,
   (ii) the optical centre,
   (iii) the focal length.
   (b) Is the lens concave, or convex?
   (c) How does the action of a diverging lens differ from that of a converging lens? Illustrate your answer with a second diagram.

2. An object of height 5 cm is placed 30 cm from the centre of a convex lens. An image is formed on a screen on the opposite side of the lens. The screen is 60 cm from the lens. By constructing a ray diagram to a suitable scale find:
   (i) the focal length of the lens; (ii) the magnification of the image.

3. A lens forms a focused image of a real object onto a screen. The separation of the screen and the object is 80 cm. The image is three times the size of the object.
   (a) What kind of lens is used?
   (b) Sketch a ray diagram of the arrangement.
   (c) Find by construction or calculation:
   (i) the distance of the image from the lens, (ii) the focal length of the lens.

4. The table below gives results obtained in an experiment to determine the focal length, f, of a converging lens: u is the object distance, and v is the corresponding image distance from the lens.

u (cm)	5.0	6.7	8.0	10.0	20.0
v (cm)	20.0	10.0	8.0	6.7	5.0

   Plot a graph of $1/u$ against $1/v$; hence determine the focal length of the lens.

5. (a) With aid of diagrams, describe two defects of vision in the human eye and show how they can be corrected.
   (b) Sketch a diagram to show the action of a magnifying glass of focal length f. Where should the image be formed for optimum viewing? What magnification is produced?

2 Uniform circular motion

Learning objectives

After completing the work in this chapter you will be able to:

1. define angular displacement and angular velocity
2. describe simple experiments to illustrate centripetal force
3. explain the applications of uniform circular motion
4. solve numerical problems involving uniform circular motion.

There are many everyday situations where objects travel in circular paths – the Earth spins on its axis and orbits the Sun; discs spin on turntables when music is being played; vehicles travel around bends in the road; electrons orbit the nucleus of atoms; the hands of a clock follow circular paths; communication satellites orbit the earth many times a day; and winds within tropical cyclones move in circular paths.

Even when an object is travelling at constant speed in a circle it is accelerating. This is because the *direction* of its velocity is changing. Newton's second law of motion tell us that a force is always needed to produce an acceleration. In this chapter you will study the nature of the centripetal forces that produce circular motion and some of their applications.

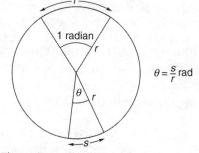

Fig. 2.1 The radian

$$\frac{2\pi r}{r} = 2\pi \text{ rad.}$$

Thus $360° = 2\pi$ rad. It follows that $180° = \pi$ rad. The following formulae can be used to convert between degree and radian measures:

$$\theta° = \frac{180}{\pi} \times \theta \text{ rad} \qquad \theta \text{ rad} = \frac{\pi}{180} \times \theta°$$

Examples

1. Express the following angles in radians:
   (a) 90°, (b) 60°.

   Answer
   (a) $90° = \frac{\pi}{180} \times 90 = \frac{\pi}{2}$ rad
   (b) $60° = \frac{\pi}{180} \times 60 = \frac{\pi}{3}$ rad

2. Express the following angles in degrees:
   (a) $\frac{\pi}{4}$ rad, (b) $\frac{2\pi}{3}$ rad

   Answer
   (a) $\frac{\pi}{4} = \frac{180}{\pi} \times \frac{\pi}{4} = 45°$
   (b) $\frac{2\pi}{3} = \frac{180}{\pi} \times \frac{2\pi}{3} = 120°$

2.1 The radian, angular displacement, angular velocity

You are familiar with the measurement of angles in degrees. A complete circle (one revolution) is divided into 360 degrees. In physics it is often convenient to use an alternative unit of angle – the **radian** (rad). Fig. 2.1 shows how the radian is defined.

One radian is the angle subtended by an arc equal in length to the radius of the circle. The angle θ is given in radians by:

$$\theta = \frac{\text{arc length}}{r} = \frac{s}{r}$$

Since the circumference of the circle is $2\pi r$, the angle in a complete circle is

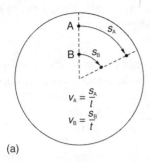

Fig. 2.2 Describing circular motion
(a) Linear displacement and speed; (b) Angular displacement and speed

Angular displacement

Fig. 2.2(a) shows two objects on a turntable. Object A is further than B from the centre of rotation and moves in a circle with a larger radius. Both objects take the same time to complete one revolution, but because object A must travel further than object B, it is travelling at a higher speed.

Fig. 2.2(b) shows an alternative way of describing the motion of the rotating objects in terms of angles. The angle θ, through which the objects have turned in time t, is called the **angular displacement**. We can see that as the objects are fixed to the same turntable, their angular displacement increases by the same amount in the same time.

Angular displacement is normally given the symbol θ. Its unit is radians.

Example

What is the increase in the angular displacement of A and B in Fig. 2.2(b) after:
(a) $\frac{1}{4}$ revolution, (b) 5 revolutions?

Answer

(a) In $\frac{1}{4}$ revolution the increase in angular displacement, $\theta = \frac{2\pi}{4} = \frac{\pi}{2}$ rad

(b) In 5 revolutions the increase in angular displacement, $\theta = 5 \times 2\pi = 10\pi$ rad

Angular velocity

Angular velocity is defined as the rate of change of angular displacement:

average angular velocity = $\dfrac{\text{increase in angular displacement}}{\text{time taken}} = \dfrac{\theta}{t}$

Angular velocity is normally given the symbol ω ('omega'). Its units are rad s^{-1}.

$\omega = \dfrac{\theta}{t}$

Example

A wheel makes 6 revolutions in 3 s. What is its angular velocity?

Answer
$\theta = 6 \times 2\pi = 12\pi$ rad; and $t = 3$ s
$\omega = \dfrac{\theta}{t} = \dfrac{12\pi}{3} = 4\pi$ rad s^{-1}

Distance travelled and angular displacement

In Fig. 2.3 the particle P has moved through an angular displacement θ. Through what distance, s, has it travelled?

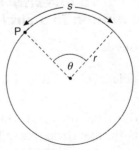

Fig. 2.3 $s = \theta r$

From the definition of the radian, $\theta = \dfrac{s}{r}$ therefore $\boxed{s = r\theta}$

Linear velocity and angular velocity

In Fig. 2.4, v is the instantaneous linear velocity of a particle moving at constant speed in a circle. The magnitude of this velocity is equal to the particle's speed.

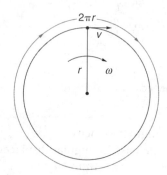

Fig. 2.4 $v = \omega r$

Consider one revolution:

$\text{speed} = \dfrac{\text{distance}}{\text{time}}$ and so $v = \dfrac{2\pi r}{t}$.

but $\omega = \dfrac{\theta}{t} = \dfrac{2\pi}{t}$ and so $\boxed{v = \omega r}$

Note: Angles must be expressed in radians for the expressions $s = \theta r$ and $v = \omega r$ to be used.

Example

A turntable with radius 15 cm is rotating at 33 rpm (revolutions per minute). Calculate:
(a) the angular velocity of the turntable,
(b) the increase in the angular displacement of a point on the rim in 10 s,
(c) the distance travelled by a point on the rim in 10 s,
(d) the speed at which a point on the rim is travelling.

Answer
(a) Number of revolutions in 60 s = 33
Therefore the increase in angular displacement in 60 s, $\theta = 33 \times 2\pi$ rad.
$\omega = \dfrac{\theta}{t} = \dfrac{33 \times 2\pi}{60} = 1.1\pi \text{ rad s}^{-1}$.

(b) $\omega = \dfrac{\theta}{t}$. Therefore $\theta = \omega t = 1.1\pi \times 10 = 11\pi$ rad.

(c) $s = \theta r = 11\pi \times 15 = 518$ cm.
(d) $v = \omega r = 1.1\pi \times 15 = 51.8$ cm s^{-1}.

Exercise 2.1

A bicycle wheel with radius 35 cm makes one revolution per second. Calculate:
(a) the angular velocity of the wheel,
(b) the increase in the angular displacement of a point on the rim in 10 s,
(c) the distance travelled by the bicycle in 10 s,
(d) the speed at which a point on the rim is travelling relative to the centre of the wheel.

2.2 Centripetal force

Newton's first law tells us that in the absence of a force an object moves at constant speed in a straight line. A force must, therefore, be acting when an object moves in a circle. In Activity 2.1 you investigate the direction and nature of this force.

Activity 2.1 To investigate the direction of the force required to produce motion in a circle at constant speed

You will need:
- string
- a mass (a large steel nut for example).

Method
Tie the mass to the string and whirl the mass in a horizontal circle (Fig. 2.5). Observe the direction of the force acting on the mass from the string.

Discussion questions
1. What kind of force does the string apply to the mass?
2. In which direction does this force act?

Fig. 2.5 Motion in a circle

The mass in Activity 2.1 is moving at a constant speed on a circular path. However, because the direction of motion of the mass is changing, it is accelerating. The force that causes its velocity to change direction is the pull from the string. This force, and the acceleration it produces, is directed towards the centre of the circle. They are called the **centripetal force** and the **centripetal acceleration**.

Fig. 2.6 shows that because the force, and therefore the acceleration it produces, is always perpendicular to the particle's instantaneous velocity its effect is to change the direction of motion without changing the speed.

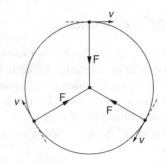

Fig. 2.6 The effect of a centripetal force

Some examples of centripetal forces are shown in Fig. 2.7.

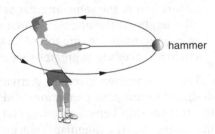

(a) The pull from the wire on a thrower's hammer

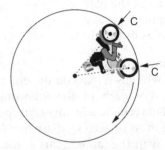

(b) The contact force from the wall on a motorcycle's wheels on the 'Wall of Death'

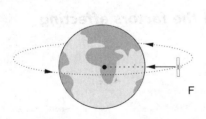

(c) The force of gravity on a satellite in orbit around the Earth

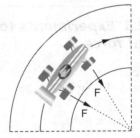

(d) The friction force on the tyres as a racing car takes a bend

Fig. 2.7 Examples of centripetal forces

Activity 2.2 Experiment to use gravity to provide a centripetal force

You will need:
- A bucket of water.

Method
Take the bucket outside into the school compound. Whirl the bucket rapidly in a vertical circle as shown in Fig. 2.8. Does the water stay in the bucket when it is upside down?

If the bucket is held upside down, the force of gravity makes the water fall vertically

(continued)

Activity 2.2 (continued)

down. But when the bucket is moving in a circle gravity helps provide the centripetal force that keeps the water moving in a circle too (the contact force with the base of the bucket also contributes to the centripetal force at higher speeds).

Fig. 2.8 Whirling a bucket

Experiments such as those in Activities 2.1 and 2.2 suggest that the size of the centripetal force required to maintain circular motion depends on the following factors:
- the mass of the object;
- the angular velocity;
- the radius of the circle.

Discussion questions

Consider whirling a mass on the end of a piece of string. For each of the following predict which would require a greater centripetal force.

1. A 1 kg mass or a 10 kg mass moving in circles with the same radius at the same angular velocity?

2. A mass moving at 1 rad s^{-1} or a similar mass moving at 2 rad s^{-1} in a circle with the same radius?

3. A mass moving in a circle with radius 1 m or a similar mass moving in a circle with radius 2 m at the same angular velocity? (Remember if the angular velocity is the same, the mass moving in the larger radius circle is travelling at greater speed.)

The experiments given in Activity 2.3 are designed to test your predictions and to reveal the relationship between centripetal force, F, the mass, m, the angular velocity, ω, and radius, r.

Activity 2.3 Experiments to investigate the factors affecting centripetal force

You will need:
- the apparatus shown in Fig. 2.9
- a stopwatch
- a partner to work with.

Basic method

1. Position the crocodile clip so that the length of the string that forms the radius is 0.5 m when the clip just touches the base of the handle.
2. Whirl the mass around, gradually increasing the speed until you can let go of the load without it falling to the ground.
3. Increase the speed until the crocodile clip just rises to the handle.
4. Your partner should time and record the number of revolutions in 60 s while you maintain a steady speed.

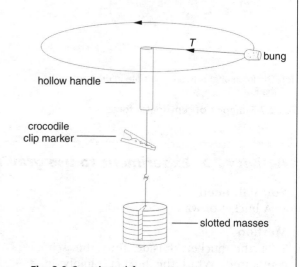

Fig. 2.9 Centripetal force apparatus

Activity 2.3 (continued)

Investigation A: the effect of angular velocity
1. Follow the basic procedure outlined above.
2. Remove/add masses to the load to change the centripetal force. Repeat the procedure and record the new number of revolutions per minute required to keep the mass moving in a circle radius r.
3. Repeat for several more loads.
4. Record your results as shown in Table 2.1.
5. Plot graphs of **(a)** ω and **(b)** ω^2 against F as shown in the sample graphs in Fig. 2.10.

Investigation B: the effect of mass
1. Follow the basic procedure outlined above.
2. Double the mass on the string.
3. Whirl the mass at the same angular velocity (your partner should use the stopwatch and count out loud at the appropriate rate to allow you to reproduce the same speed).
4. Remove/add masses to the load to change the centripetal force until the new mass whirls in a circle of the same radius at the same rate as the smaller mass.
5. Repeat for several more whirling masses.
6. Record your results as shown in Table 2.2.
7. Plot a graph of m against F as shown in the sample graph in Fig. 2.11 on page 28.

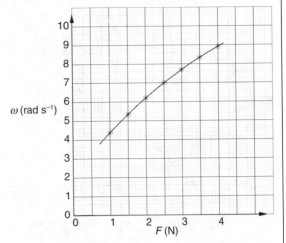

(a) Graph of angular velocity (ω) against centripetal force (F) for $m = 0.1$ kg and $r = 0.5$ m

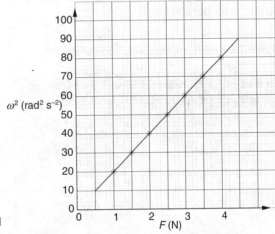

(b) Graph of ω^2 against F for $m = 0.1$ kg and $r = 0.5$ m

Fig. 2.10 Sample graphs for Investigation A
(a) graph of angular velocity (ω) against centripetal force (F)
(b) graph of ω^2 against F

Table 2.1 Radius, r = constant; mass, m = constant

Load, L (kg)	Centripetal force, $F = Lg$ (N)	Number of revolutions per minute, n	$\omega = \frac{n \times 2\pi}{60}$ (rad s^{-1})	ω^2
0.10	1.0	43	4.5	20.3
0.15	1.5	52	5.4	29.2
0.20	2.0	60		
0.25				
0.30				

(continued)

Activity 2.3 (continued)

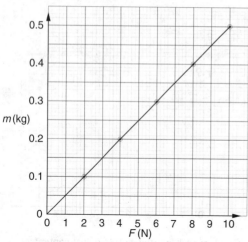

Graph of mass (*m*) against centripetal force (*F*) for r = 0.5 m and ω = 2π

Fig. 2.11 Sample graph for Investigation B: mass (*m*) against centripetal force (*F*)

Table 2.2 Radius, *r* = constant; angular velocity, ω = constant

Load, L (kg)	Centripetal force, F = Lg (N)	Mass, m (kg)
0.20	2.0	0.1
0.40	4.0	0.2
0.60	6.0	0.3

Investigation C: the effect of radius
1. Follow the basic procedure outlined above.
2. Move the crocodile clip to increase the radius of the circle by 10 cm.
3. Whirl the mass at the same angular velocity (your partner should use the stopwatch and count out loud at the appropriate rate to allow you to reproduce the same speed).
4. Remove/add masses to the load to change the centripetal force until the mass whirls in a circle of increased radius at the same rate as the smaller circle.

Table 2.3 Mass, *m* = constant; angular velocity, ω = constant

Load, L (kg)	Centripetal force, F = Lg (N)	Radius, r (m)
0.12	1.2	0.3
0.16	1.6	0.4
0.20	2.0	0.5

5. Repeat for several more radius values.
6. Record your results as shown in Table 2.3.
7. Plot a graph of *r* against *F* as shown in the sample graph in Fig. 2.12.

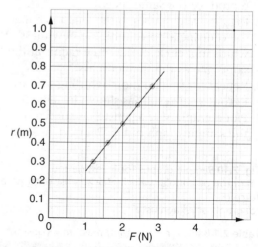

Graph of radius (*r*) and centripetal force (*F*) for m = 0.1 kg and ω = 5π rads^{-1}

Fig. 2.12 Sample graph for Investigation C: radius (*r*) against centripetal force (*F*)

Discussion
The investigation of the centripetal force, *F*, acting on a mass, *m*, moving in a circle of radius *r* at angular velocity ω shows that:
- $F \propto m$
- $F \propto \omega^2$
- $F \propto r$.

Experiment and theory confirm that if all quantities are expressed in SI units, then the centripetal force is given by:

$$F = m\omega^2 r$$

Since $v = \omega r$, and therefore $\omega = \frac{v}{r}$, the centripetal force is also given by:

$$F = \frac{mv^2}{r}$$

2.3 Examples and applications of uniform circular motion

Fig. 2.13 shows two examples of circular motion. Each is discussed in turn in the text that follows.

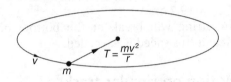

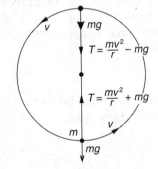

Fig. 2.13 Applications of $F = m\omega^2 r$ and $\frac{mv^2}{r}$

Motion in a horizontal circle

For a mass whirled in a horizontal circle, the centripetal force is provided by the tension in the string.

$$T = m\omega^2 r = \frac{mv^2}{r}$$

Fig. 2.14 shows the path the mass would follow if the string were to break. It flies off in a straight line in the direction it was moving at the point the string parted – at a tangent to a circle.

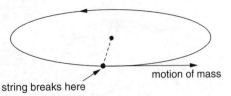

Fig. 2.14 What happens when the string breaks?

Note: In practice the tension in the string must support the weight of the mass as well as provide the centripetal force. The string, therefore, must be inclined to the horizontal so that its tension has a vertical component equal to mg. However, if the object is whirled rapidly the angle of inclination becomes very small and, to a good approximation, $T = m\omega^2 r$.

Motion in a vertical circle

For a mass whirled in a vertical circle at constant speed the tension in the string varies. At the bottom of the circle (as Fig. 2.13(b) shows) the tension must provide the centripetal force and support the weight of the mass:

$$T_{\text{bottom}} = \frac{mv^2}{r} + mg \qquad (1)$$

At the top of the circle the weight provides part of the centripetal force and the tension therefore reduces:

$$\frac{mv^2}{r} = T_{\text{top}} + mg$$

$$T_{\text{top}} = \frac{mv^2}{r} - mg \qquad (2)$$

Equation 2 shows that the tension in the string drops to zero when $v = \sqrt{rg}$.

At this speed the weight of the mass provides all the centripetal force to maintain motion in a circle as the mass moves over the top of the circle. This is the minimum speed required to maintain vertical motion. If the speed falls below this value the string goes slack and the stone drops out of its circular path as it reaches the top of its motion. Try it yourself with a mass on a string – a minimum speed is required before a vertical circle can be produced. Two related examples are shown in Fig 2.15.

For the whirling bucket and the motorcycle stunt, the centripetal force at the top of the motion is provided by the weight plus the total contact force between the bucket/wall and the water/motorcycle.

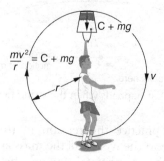
(a) The whirling bucket

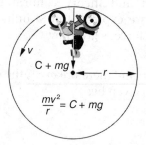
(b) The motorcycle stunt

Fig. 2.15 A minimum speed is required to perform these tricks

Now, $\dfrac{mv^2}{r} = C + mg$

The contact force is thus given by

$C = \dfrac{mv^2}{r} - mg$

and this becomes zero when $v = \sqrt{rg}$.
The water/motorcycle will thus fall out of its circular path if the speed drops below this value.

Example

A 0.5 kg mass is swung in a vertical circle of radius 1 m on a string.
(a) Calculate the minimum speed required for the circle to be maintained.
(b) If the speed falls below this value, where will the string first go slack?
(c) If the breaking tension of the string is 200 N, at what speed can the mass be rotated before the string breaks? Where in the circle will the string break?
(Take $g = 10$ m s^{-2})

Answers
$m = 0.5$ kg, $r = 1$ m and $g = 10$ m s^{-2}.
(a) The minimum speed required to maintain vertical motion is given by:
$v = \sqrt{rg} = \sqrt{1 \times 10} = 3.2$ m s^{-1}
(b) If the speed falls below this value, the string will go slack at the top of the circle.
(c) Tension is greatest at the bottom of the circle – when the tension reaches 200 N the string will snap.

$T_{\text{bottom}} = \dfrac{mv^2}{r} + mg = 0.5v^2 + 5 = 200$ N

$0.5v^2 = 195$
$v^2 = 390$
$v = 19.7$ m s^{-1}

The string will break at the bottom of the circle if this speed is exceeded.

Motion on circular tracks

Fig. 2.16 shows a car of mass m taking a circular bend of radius r at a speed v. The centripetal force required to keep the car on its path is

$\dfrac{mv^2}{r}$

This force is provided by friction between the car's tyres and the road.

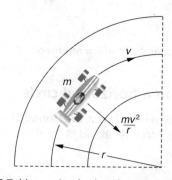
Fig. 2.16 Taking a circular bend

In Form Three you learned that friction has a maximum, or limiting, value, F_{max} determined by the nature of the surfaces, and the contact force, C, between them.

In this case, the contact force is equal to the weight of the car, *mg*, which is fixed.

If the force tending to produce relative motion of the surfaces exceeds F_{max}, then they will slide over each other. Thus, the maximum speed, v_{max}, at which the car can take the bend without skidding may be found from

$$\frac{mv_{max}^2}{r} = F_{max}$$

where *r* is the radius of curvature of the bend.

If the speed of the vehicle exceeds this value it slips or skids otwards. The friction between car tyres and the road surface is much higher on a dry and firm surface than on wet or loose surfaces. It is therefore necessary to take bends at lower speeds on a wet road or a poor surface to avoid skidding.

Banking

Race tracks for cars and athletes are often banked at the bends. This enables the bends to be taken at higher speeds without skidding or slipping. The action of banking is shown in Fig. 2.17. The horizontal component of the contact force of the track on the wheels (or on an athlete's feet) is now able to contribute to the centripetal force so the bend can be taken a greater speed without skidding.

For a given angle of banking there is an ideal speed at which the bend should be taken. If the speed is below the ideal then friction is necessary to stop movement down the banking. If the ideal speed is exceeded,

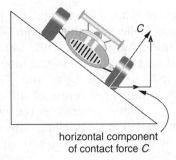

Fig. 2.17 A banked track

friction stops movement outwards and upwards. At the ideal speed there is no tendency to move sideways and no sideways friction is required, either up or down. The bend can be taken at this speed even if the surface is perfectly smooth. On many tracks, the banking is 'saucer shaped' – the angle of banking increases with height. Faster moving athletes or vehicles can thus take the bend higher up the track where the banking is steeper.

The centrifuge

A centrifuge is used to separate solids suspended in liquids, immiscible liquids of different densities, or gas samples of different density.

Imagine a tube containing fine grains of a solid suspended in water, hanging on a string. Now imagine swinging the tube rapidly in a

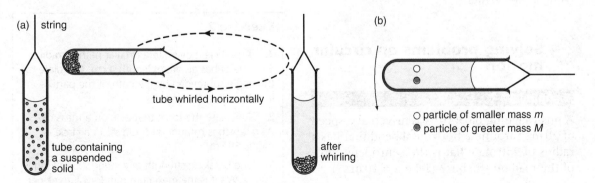

Fig. 2.18 The action of a centrifuge
(a) The principle of the centrifuge
(b) The behaviour of particles in a centrifuge

horizontal circle. After a little while, the more dense material collects at the bottom of the tube. To understand why this happens remember that, at any point in the tube, the force needed to accelerate the liquid continuously towards the centre of the circle is given by $F = m\omega^2 r$, where r is the radius along which that point is travelling. This force is provided by the pressure of the surrounding liquid.

Now consider a particle of liquid (Fig. 2.18(b)) replaced by a particle of solid of greater density. The force from the surrounding liquid remains the same but, since m is greater, this force is no longer sufficient to keep the particle moving in its circular path and it will therefore move outwards towards the base of the tube – the greater the density of the particle, the more rapid its motion through the liquid. In a similar way, particles less dense than the liquid will experience a force greater than that required to keep them in a circular path and will therefore move inwards towards the surface of the liquid.

The overall effect of centrifugation is to separate particles by density. Materials of higher density collect at the bottom of the tube, materials of lower density at the top.

If the tube is whirled very rapidly, the end of the tube might not be strong enough and the tube will break open because of the high reaction force exerted on it by the contents. Centrifuge tubes used in laboratories are therefore made especially strong. The technique provides a rapid method of separating solids and liquids. It can be much faster than using a filter paper.

2.4 Solving problems on circular motion

Example 1

A bus weighing 6.0×10^4 N is driven at a speed of 20 m s^{-1} around a horizontal bend that has a radius of 250 m. What is the centripetal force of the road on the bus? (Take $g = 10 \text{ m s}^{-2}$)

Answer

$v = 20 \text{ m s}^{-1}$ and $r = 250 \text{ m}$.
$W = mg$

So:
$m = \dfrac{W}{g} = \dfrac{6.0 \times 10^4}{10} = 6000 \text{ kg}$

$F = \dfrac{mv^2}{r} = \dfrac{6000 \times 20^2}{250} = 9600 \text{ N}$

Example 2

A person of mass 50 kg is on a swing which has a speed of 10 m s^{-1} at the lowest point of its motion. The ropes of the swing are 2.5 m long. Find the tension in the ropes. Assume $g = 10 \text{ m s}^{-2}$.

Answer

$m = 50 \text{ kg}$, $v = 10 \text{ m s}^{-1}$ and $r = 2.5 \text{ m}$.
At the bottom of the motion
$T = \dfrac{mv^2}{r} + mg = \dfrac{50 \times 10^2}{2.5} + 50 \times 10$
$= 2000 + 500$
$= 2500 \text{ N}$

Example 3

A train is travelling on a track, which is part of a circle of radius 600 m, at a constant speed of 50 m s^{-1}.
(a) What is the angular velocity of the train?
(b) What centripetal force acts on the train if its mass is 2000 kg?

Answer

$r = 600 \text{ m}$, $v = 50 \text{ m s}^{-1}$ and $m = 2000 \text{ kg}$.
(a) $v = \omega r$, therefore
$\omega = \dfrac{v}{r} = \dfrac{50}{600} = 0.083 \text{ rad s}^{-1}$

(b) $F = m\omega^2 r$
$= 2000 \times \left(\dfrac{50}{600}\right)^2 \times 600 = 8333 \text{ N}$

Exercise 2.2

1. A particle moving in a circular path of radius 5 cm describes an arc of length 4 cm. Calculate the angle subtended by the path of the particle in (a) radians, (b) degrees.

2. Calculate the force that acts on a mass of 2 kg which is rotating at 5 rad s^{-1} in a circle of radius 50 cm.

3. The breaking strength of a string 2.5 m long is 100 N. What is the maximum rpm (revolutions per minute) at which the string can retain a 2 kg mass attached to its end?

(continued)

Exercise 2.2 (continued)

4. A crane lifts a load at a speed of 5.4 m s^{-1}. The drum onto which the wire is wound has a diameter of 36 cm. Calculate the angular velocity of the drum.

5. Fig. 2.19 shows a ball of mass 0.30 kg swung in a vertical circular path on the end of a string of length 80 cm so that it moves with a constant speed of 4.0 m s^{-1}. Which of the following statements about the motion is false?

(a) The tension in the string is greater at Q than at P.

(b) The net force on the ball is constant in magnitude.

(c) If the string breaks, the body will move radially outwards.

(d) The net force on the body is always directed towards the centre of the circle.

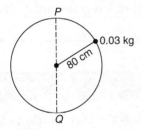

Fig. 2.19

6. An apple is whirled round a horizontal circle on the end of a string which is tied to the stalk. It is whirled faster and faster and at a certain speed the apple is torn from the stalk. Why is this?

7. Explain why it is easier to ride a motorcycle around a bend in the road if the surface is dry rather than if it is wet.

8. (a) What is 'uniform circular motion'?

(b) What force maintains it?

(c) What factors determine the magnitude of a centripetal force?

9. A motorcycle is travelling at a constant speed of 72 km h^{-1} around a circular track of radius 150 m.

(a) How long does the motorcyclist take to complete one full circuit of the track?

(b) What is his angular velocity?

(c) If the total mass of the motorcycle and rider is 250 kg what is the centripetal force?

10. The 'Wall of Death' is a stunt ride in which a motorcyclist rides his machine around the inside of a large cylinder, as shown in Fig. 2.20. The motorcycle and its rider have a total mass of 200 kg. The wall has a radius of 5.0 m and the motorcycle is moving at a constant speed of 10 m s^{-1} at the top of the loop.

(a) Find the size of the force the wall exerts on the rider and bicycle.

(b) Why does the rider not fall to the bottom of the cylinder?

(c) What is the minimum speed of the motorcycle to complete this stunt?

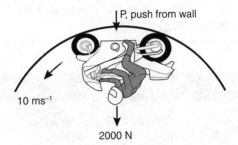

Fig. 2.20 The 'Wall of Death'

11. A car of mass 1200 kg is moving with a velocity of 25 m s^{-1} around a horizontal bend of radius 150 m. Determine the minimum frictional force between the tyres and the road that will prevent the car from sliding on the road surface.

12. An object of mass 0.5 kg attached to the end of a string is whirled round in a horizontal circle of radius 2 m. The string breaks when the tension in it exceeds 100 N.

(a) Calculate the maximum angular velocity of the object.

(b) The object is now whirled in a vertical circle by the string, increasing its speed from zero. Will the string break when the object is at the top or the bottom of the circle? Give reasons for your answer.

Summary

- One radian is the angle subtended by an arc equal in length to the radius of the circle. The angle θ is given in radians by:
$\theta = \dfrac{\text{arc length}}{r}$ rad.
- Angular displacement, $\theta = \dfrac{s}{r}$.
- Angular velocity, $\omega = \dfrac{\theta}{t}$.
- $v = \omega r$.
- A centripetal force F is required to maintain uniform circular motion:
$F = m\omega^2 r$,
$F = \dfrac{mv^2}{r}$.
- For a mass whirled in a horizontal circle the centripetal force is provided by the tension in the string:
$T = m\omega^2 r = \dfrac{mv^2}{r}$.
- For a mass whirled in a vertical circle at constant speed, the tension in the string varies:
$T_{\text{bottom}} = \dfrac{mv^2}{r} + mg$,
$T_{\text{top}} = \dfrac{mv^2}{r} - mg$.
- The minimum speed to maintain motion in a vertical circle is given by $v = \sqrt{rg}$.
- Banking enables bends to be taken at higher speeds without skidding or slipping.
- The effect of centrifugation is to separate particles by density – materials of higher density collect at the bottom of the tube, materials of lower density at the top.

Review questions

1. A car travelling in a circular track of radius 50 metres at constant speed completes 6 revolutions per minute.
   (a) What is the period of motion (time taken for one revolution)?
   (b) What is the magnitude of the car's velocity at any instant?
   (c) What is the magnitude of the centripetal acceleration?
   (d) How many radians are there in one revolution?
   (e) What is the angular velocity of the car?

2. (a) A person in a car rounding a sharp curve must hold onto the seat, or they will slide outwards. What force are they providing by holding the seat? Why is this force necessary?
   (b) Explain why race tracks and railways are often banked on curves.

3. A satellite orbiting the earth takes 27 days to make one revolution around the Earth. Assume that its orbit is circular with radius 3.8×10^5 km.
   (a) What is the acceleration of the satellite?
   (b) The satellite has mass of 7.4×10^{22} kg; what is the centripetal force?
   (c) What provides this force?
   (d) What is the reaction force on the Earth? Where does it act?
   (e) The Earth has mass 5.98×10^{24} kg; what is the acceleration of the Earth towards the satellite?

4. (a) A stone of mass m at the end of rope is whirled in a vertical circle of a radius r at constant speed. Show that the speed at which the string would just become slack at the highest point is equal to $\sqrt{gr}$, where g is the acceleration due to gravity.
   (b) A coin is placed on the turntable of a record player and the speed of the turntable is increased. What force keeps the coin moving in a circle with the turntable? Discuss the factors that determine the rotation speed at which the coin starts to slide.

3 Floating and sinking

Learning objectives

After completing the work in this chapter you will be able to:

1. state Archimedes' principle
2. verify Archimedes' principle
3. state the law of flotation
4. define relative density
5. describe the applications of Archimedes' principle and relative density
6. solve numerical problems involving Archimedes' principle.

3.1 Archimedes' principle and the law of flotation

If you push a football under water in a pool or tank, it displaces (pushes aside) the water from the space it occupies. Gravity tries to pull the surrounding water back into the space from which it has been displaced, and the object experiences an upward force, or **upthrust**, as a result of the water pressure – the more of the ball you submerge, the greater the upthrust. The Greek philosopher and scientist Archimedes investigated the relationship between the upthrust and the displacement. His findings explain why some objects sink, while others float.

In Activity 3.1 you will repeat Archimedes' investigations.

Activity 3.1 Experiment to investigate the upthrust on a submerged object

You will need:
- a sensitive newton meter
- a regular object such as a cuboidal stone or metal block
- some thread
- a large Eureka can (or measuring cylinder)
- a beaker
- a balance
- a stand.

Method
1. Set up the apparatus as shown in Fig. 3.1. Fill the Eureka can to the level of the spout with water.
2. Record the weight of the empty beaker.
3. Record the weight of the object in air.
4. Make sure the beaker is positioned to collect the water that overflows from the Eureka can.
5. Lower the object into the can until it is approximately one-fifth submerged.
6. Measure and record the apparent weight of the object and the weight of the water displaced.
7. Lower the object further into the water in stages and record the weight readings at each step until the object is fully submerged, as shown in Table 3.1.
8. Plot a graph of the weight of water displaced against the apparent loss in weight of the object. Fig. 3.2 shows an example of the results obtained.

(continued)

Activity 3.1 (continued)

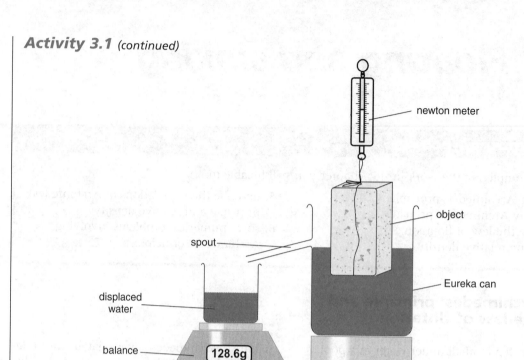

Fig. 3.1 Experimental arrangement

Table 3.1 Weight of object in air = ... N

Reading on newton meter (apparent weight of object) (N)	Loss in weight of object (N)	Weight of water displaced (N)

Discussion questions
1. Why does the object appear to lose weight when it is submerged?
2. How do your results indicate that the loss in weight is related to the properties of the displaced water?

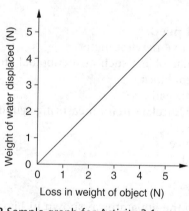

Fig. 3.2 Sample graph for Activity 3.1

Archimedes' principle

Through experiments such as Activity 3.1 Archimedes formulated his principle.

When an object is wholly or partially immersed in a fluid, the upthrust on the object is equal to the weight of fluid displaced.

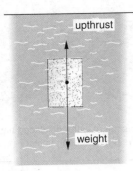

Fig. 3.3 Apparent weight loss

The upthrust acts in the opposite direction to the weight. The apparent loss in weight of the sample is thus equal to the upthrust.

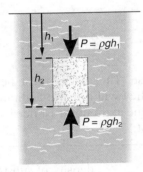

Fig. 3.4 Explanation of upthrust

Upthrust can be explained by the pressure difference between the top and the bottom of the object. As you learned in Form Two, the pressure in a liquid increases with depth ($P = \rho gh$). The upward pressure on the base of the object is thus greater than the downward pressure on its surface. The pressure difference produces a net upward force.

Fig. 3.5(a) shows a 'cube' of fluid in a vessel. The cube of fluid shown is stationary so the upthrust from the surrounding fluid on the cube must exactly balance its weight (mg) – the upthrust is equal to the weight of the fluid in the cube.

Fig. 3.5(b) shows the fluid cube replaced by a solid sample of the same size. The *surrounding* fluid has not changed so the upthrust must

still be the same. The upthrust on the submerged solid is thus equal to the weight of the fluid it has displaced.

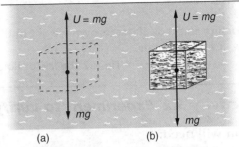

Fig. 3.5 Explanation of Archimedes' principle
(a) Cube of fluid
(b) Cube of a solid of the same volume

Floating and sinking

Consider the forces acting on the samples A and B in Fig. 3.6. Sample A is a material with a greater density than water, so its weight is greater than that of an equal volume of water. The upthrust on it is equal to the weight of the water displaced. This is insufficient to balance its weight (mg) so, when it is released, sample A sinks.

Sample B is made from a material which is less dense than water (cork for example). The upthrust on it is greater than its own weight (mg) so, when sample B is released, it rises to the surface.

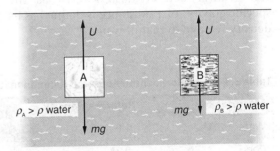

Fig. 3.6 Sample A sinks, sample B rises

Law of flotation

A sample which is less dense than water floats at the water surface partially submerged. Since the sample is in equilibrium we can deduce that the net force acting on it is zero – the upthrust is equal and opposite to the weight. Consideration of Archimedes' principle (the upthrust is equal to the weight of the fluid displaced) gives rise to the law of flotation which states:

A floating object displaces its own weight of the fluid in which it floats.

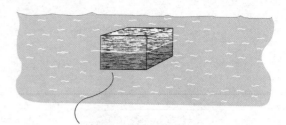

Fig. 3.7 The law of flotation

weight of water displaced equals weight of floating object

Activity 3.2 Experiment to verify the principle of flotation

You will need:
- Eureka can
- different types of wood blocks
- balance.

Method
1. Weigh and record the weight of the block.
2. Fill the Eureka can to the spout with water.
3. Slowly lower the block into the water until it floats.
4. Collect and weigh the water displaced.
5. Repeat the experiment with different samples.
6. Record your results as shown in Table 3.2.

Table 3.2

Weight of block (N)					
Weight of water displaced (N)					

Discussion
To confirm the principle of flotation you should observe that the weight of the water displaced is equal to the weight of the block in each case. The blocks must be lowered slowly or they will displace additional water to provide the extra upthrust needed to bring them to rest.

3.2 Relative density

The relative density (also known as the specific gravity) of a substance is defined as its density relative to that of water.

$$\text{Relative density} = \frac{\text{density of substance}}{\text{density of water}}$$

$$= \frac{\text{mass of substance}}{\text{mass of an equal volume of water}}$$

The relative densities of some materials are listed in Table 3.3.

Table 3.3 The relative densities of some common materials

Material	Relative density	Material	Relative density
Water (at 4°C)	1.0	Paraffin	0.8
Cork	0.25	Glass	2.5
Gold	19.3	Iron	7.9
Aluminium	2.7	Brass (60/40)	8.4
Mercury	13.5	Polythene	0.95
Lead	11.3	Cedar wood	0.55
Granite	2.7		

Examples

Use the law of flotation to explain the following facts.

1. A solid will float on the surface of a liquid if the relative density of the solid is less than that of the liquid.
2. A solid sinks in a liquid if its relative density is greater than that of the liquid.
3. The lower the relative density of the liquid, the greater the fraction of the solid that must be submerged before the upthrust is sufficient to make the solid float.
4. The higher the relative density of the liquid, the more of the solid is above the surface when it floats.

These facts are illustrated in Fig. 3.8.

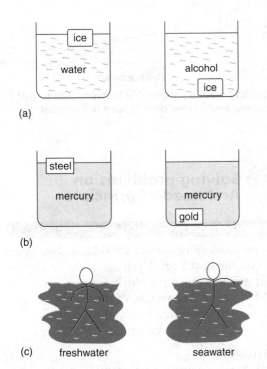

Fig. 3.8 Relative density and flotation
(a) Ice (relative density 0.92) floats on water (relative density 1.0) but sinks in alcohol (relative density 0.8).
(b) Steel (relative density 7.9) floats on mercury (relative density 13.5), but gold (relative density 19.3) sinks in mercury.
(c) A human body floats higher in salt water than in fresh water (this is why it is easier to swim in the sea than in a freshwater lake).

3.3 Applications of Archimedes' principle and relative density

Archimedes' principle has many important practical applications. These include:
- ship design;
- the operation of balloons;
- the construction of hydrometers.

Ships

A ship floats at different levels depending on the density of the water it is in. It floats lower in fresh river water than in seawater, and lower in warm water than in cold water. Ships have a special safety mark painted on their side to indicate how heavily the ship should be loaded in different kinds of water – the mark is called the **Plimsoll line**. Samuel Plimsoll was a member of the British parliament and he introduced the marking in 1885 to protect sailors from ship owners who loaded so much cargo onto their ships that the vessels were likely to sink.

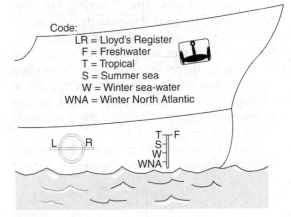

Fig. 3.9 The Plimsoll line on a ship

Balloons

An object in air also experiences an upthrust. The upthrust is equal to the weight of the air displaced. If the upthrust is greater than the weight of the object, the object will rise. This is the principle of balloons filled with hydrogen, helium or hot air. They rise because hydrogen, helium and hot air are all less dense than air.

The hydrometer

A hydrometer is an instrument for measuring the relative densities of liquids – for example

Chapter 3: Floating and sinking

Fig. 3.10 A hot air balloon

battery acid, fuels, or water/alcohol mixtures.

Fig. 3.11 shows the principle of the hydrometer. The sealed glass body sinks in the liquid until the weight of the liquid displaced is equal to the weight of the hydrometer. The hydrometer, therefore, sinks more deeply in a liquid of low relative density than in a liquid of high relative density. The stem of the hydrometer is narrow to make the instrument more sensitive.

Once the main bulb is fully submerged a further increase in depth produces only a small increase in the total submerged volume. The relative density divisions are therefore more widely spaced, and can be read more accurately than would be the case if the stem had a larger cross-section. Note that the scale divisions are not evenly spaced (the scale is not linear). This is because the depth to which the hydrometer sinks *decreases* as the relative density of the liquid *increases* (the two quantities are related inversely as illustrated in Fig. 3.12); a greater volume of the hydrometer must be submerged to displace its own weight of a liquid of lower relative density.

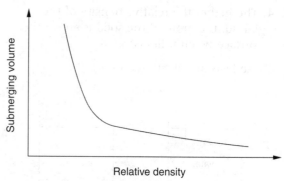

Fig. 3.12 Inverse relationship between submerged volume and relative density for a hydrometer

3.4 Solving problems on Archimedes' principle

Example 1

A metal block of density 2500 kg m^{-3} occupies a volume of 0.1 m^3. Find:
(a) the weight of the block,
(b) the apparent weight of the block when immersed in water of density 1000 kg m^{-3}. (Assume $g = 10$ m s^{-2})

Answer
(a) Mass = density × volume
   = 2500 × 0.1 = 250 kg
   weight = mg = 250 × 10 = 2500 N
(b) Apparent weight loss = upthrust on metal block = weight of water displaced
   = volume × density of water × g
   = 0.1 × 1000 × 10 = 1000 N
   Apparent weight of metal block in water
   = 2500 − 1000 = 1500 N

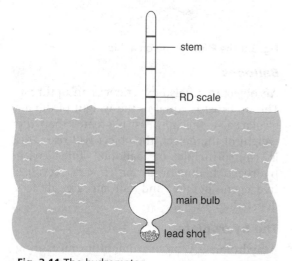

Fig. 3.11 The hydrometer

Example 2

A solid object floats in water with 30% of its volume above the surface. Find the density of the solid.
(The density of water = 1000 kg m^{-3})

Answer

Let the mass, volume and density of the object be m, V and ρ respectively.

From the principle of flotation, the weight of water displaced is equal to the weight of the body. The mass of the water displaced is therefore also m.

The volume of water displaced = $0.7V$

The density of water = 1000 kg m^{-3}

Now, $m = \rho V = 1000 \times 0.7V$

Therefore, $\rho = 1000 \times 0.7$ kg m^{-3} = 700 kg m^{-3}

Example 3

A large balloon is filled with hot air to a volume of 200 m³ (Fig. 3.13). It has a total weight of 2200 N. It is held to the ground by a vertical rope. Given the density of air is 1.2 kg m^{-3} and that $g = 10$ m s^{-2} find:
(a) the upthrust acting on the balloon,
(b) the tension in the rope.

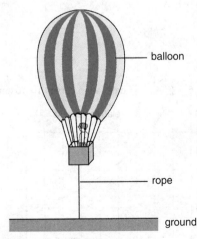

Fig. 3.13 Hot air balloon

Answer
(a) Upthrust = weight of air displaced
 mass of air = density of air × volume of air
 = 1.2 × 200
 = 240 kg
 weight of displaced air = 240 × g = 2400 N
 Therefore upthrust = 2400 N

(b) Since the balloon is stationary the tension in the rope must balance the resultant of the upthrust and the balloon's weight.
Tension = 2400 − 2200 = 200 N

Exercise 3.1

1. Explain why a ship floats higher in salty water than in fresh water.

2. A student drops a piece of wood with volume 24 cm³ into a measuring cylinder containing 60 cm³ of water. The water level rises to 72 cm³.
(a) Does the wood float or sink?
(b) What is the density of the wood?

3. An object weighs 60 N in air and 40 N when totally immersed in water.
Calculate:
(a) the upthrust on the object in water,
(b) the weight of the water displaced,
(c) the mass of water displaced,
(d) the volume of water displaced,
(e) the volume of the object,
(f) the mass of the object,
(g) the density of the object.

4. A rubber ball floats in water with 40% of its volume above water level. Calculate the density of the ball.

3.5 Project to construct a hydrometer

Fig. 3.14 shows a design for a simple hydrometer. The special feature of this hydrometer is that it has uniform cross-section, A, and can therefore be calibrated easily.

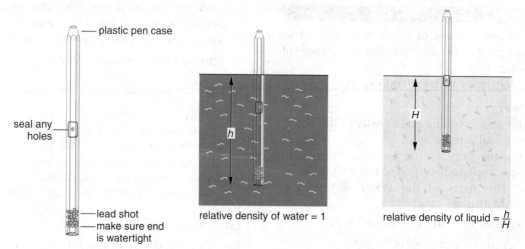

Fig. 3.14 A simple hydrometer

Activity 3.3 Making a simple hydrometer

Calibrate the hydrometer as follows.
1. Load the hydrometer with lead shot so that it floats in water, approximately two-thirds submerged. Let the mass of the hydrometer = m; the cross-section area = A.
2. Record the depth, h, to which the hydrometer sinks in pure water.
   Volume of water displaced = Ah
   Mass of water displaced = $\rho_w Ah = m$,
   where ρ_w is the density of water.

Now let H be the depth to which the hydrometer sinks in another liquid of density ρ.
Mass of liquid displaced = $\rho AH = m$
Therefore $\rho AH = \rho_w Ah$
So the relative density of the liquid
$= \dfrac{\rho}{\rho_w} = \dfrac{h}{H}$

3. Use your measurement of h to construct a calibration chart as illustrated in the example that follows. Suppose $h = 10$ cm. Then
   relative density $= \dfrac{10}{H}$
A graph of relative density against H, plotted using this equation, is shown in Fig. 3.15.

4. The hydrometer can now be used by noting the depth to which it sinks in a liquid and reading the relative density from the chart. Use your hydrometer to measure the relative densities of various liquids including paraffin, cooking oil and brine.

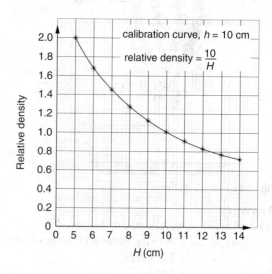

Fig. 3.15 Hydrometer calibration for $h = 10$ cm

Summary

- Archimedes' principle states that when an object is wholly or partially immersed in a fluid, the upthrust on the object is equal to the weight of fluid displaced.
- The upthrust acts in the opposite direction to the weight – the apparent loss in weight of the object is equal to the upthrust.
- An object which is less dense than water floats at the water surface, partially submerged.
- The law of flotation states that a floating object displaces its own weight of the fluid in which it floats.
- Relative density = $\dfrac{\text{density of substance}}{\text{density of water}}$

  = $\dfrac{\text{mass of substance}}{\text{mass of an equal volume of water}}$.
- A hydrometer is an instrument for measuring the relative densities of liquids, for example battery acid, fuels, or water/alcohol mixtures.

Review questions

1. **(a)** State Archimedes' principle.
   (b) A block of wood weighing 2.0 N is held under water by a string attached to the bottom of a container. The tension in the string is 0.5 N. Determine the density of the wood.

2. A solid displaces 11.0 cm³ of paraffin when floating and 40.0 cm³ when fully immersed in it. Given that the density of paraffin is 0.80 cm³, calculate the density of the solid.

3. A rectangular metal bar, of dimensions 0.02 m × 0.02 m × 0.2 m and a density 2700 kg m⁻³, is supported inside a liquid of density 800 kg m⁻³ by a thread which is attached to a spring balance. The long side is vertical and the surface is 0.1 m. below the surface of the liquid.
   (a) Calculate the force due to the liquid on:
   (i) the lower surface of the bar,
   (ii) the upper surface of the bar.
   (g = 10 m s⁻²)
   (b) Calculate the upthrust, and hence determine the reading on the balance.

4. Explain why a solid block of steel sinks, but a ship built from steel floats.

4 Electromagnetic spectrum

Learning objectives

After completing the work in this chapter you will be able to:

1. describe the complete electromagnetic spectrum
2. state the properties of electromagnetic waves
3. describe the methods of detecting electromagnetic radiations
4. describe the applications of electromagnetic radiations
5. solve numerical problems involving $c = f\lambda$.

4.1 Electromagnetic spectrum

Light travels as a transverse wave, like a water wave, but there is a major difference. Unlike water waves and sound waves, light waves do not need a medium in order to travel. Light consists of vibrating waves of electricity and magnetism which can carry energy through the vacuum of empty space. Light is an **electromagnetic wave**.

Light is not the only kind of electromagnetic wave. In fact it forms just one narrow band of the electromagnetic spectrum. The complete spectrum is shown Fig. 4.1, along with some of the sources and detectors of the different radiation bands. All electromagnetic waves travel and carry energy in the same way as light does, but electromagnetic waves with different wavelengths are given different names because of the different ways they are produced or detected, and the different effects they have on matter.

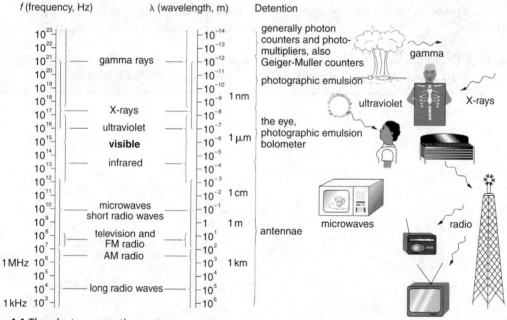

Fig. 4.1 The electromagnetic spectrum

Activity 4.1 Investigating sources and detectors of electromagnetic radiation

Make a display of sources and detectors of electromagnetic radiation in your classroom. These could include:
- a radio receiver with an aerial
- a torch
- an electric heater
- a microwave cooker
- photographic film
- a photographic exposure meter
- photovoltaic cells (solar cells)
- fluorescent dyes.

Write labels for each source/detector with details of the radiation it produces or responds to.

4.2 The properties of electromagnetic waves

Fig. 4.2 shows an electromagnetic wave. It consists of vibrating electric and magnetic fields at right angles to each other – these carry energy as they travel through space. All electromagnetic radiation travels with the same speed through a vacuum, the speed of light. The speed of light, c, in a vacuum is 3×10^8 m s^{-1}.

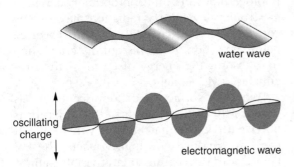

Fig. 4.2 An electromagnetic wave

The wave equation $v = f\lambda$, which you first met in Form Two, applies to electromagnetic waves as it does to all wave motions. For electromagnetic waves travelling in a vacuum it is written:

$c = f\lambda$,

where c is the speed of light in a vacuum.

Since c is a constant, with the same value for all electromagnetic waves, the wave equation shows that the wavelength is inversely proportional to the frequency:

$\lambda = c/f$

Thus, long wavelength electromagnetic waves have low frequencies, while shorter wavelengths have higher frequencies.

Example

The wavelengths of visible light are between 4×10^{-7} m (blue) and 7×10^{-7} m (red). Calculate the frequencies of (a) blue, and (b) red light. The speed of light is 3×10^8 m s^{-1}.

Answer
Using $c = f\lambda$

(a) $f_{blue} = \dfrac{3 \times 10^8}{4 \times 10^{-7}} = 7.5 \times 10^{14}$ Hz

(b) $f_{red} = \dfrac{3 \times 10^8}{7 \times 10^{-7}} = 4.3 \times 10^{14}$ Hz

Electromagnetic radiation with a wavelength shorter than 4×10^{-7} m or longer than 7×10^{-7} m is not visible to the human eye.

The energy carried by electromagnetic waves depends on their frequency. The radiations at the high frequency/short wavelength end of the spectrum (gamma and X-rays for example) carry far more energy than radiations at the low frequency/long wavelength end of the spectrum (radio waves for example).

4.3 The detection of electromagnetic radiations

Visible light is the only form of radiation we can see with our eyes. We can detect the presence of infrared radiation by the heat it generates in our skin, but we cannot generally sense radiations from the other regions of the electromagnetic spectrum. We can, however, detect electromagnetic radiation of all wave-lengths with specially designed sensors and instruments.

Depending on the wavelength, detection devices for electromagnetic radiation make use of electrical effects, chemical changes or temperature changes produced when the energy of the radiation is absorbed by matter. Specific types of device are discussed in turn below.

Radio waves and microwaves

Radio waves and microwaves are detected by aerials or antennae. The energy of these wave bands is absorbed by the conduction electrons in metals, causing the electrons to vibrate at the same frequencies as the waves. The resulting alternating current can be amplified electronically to produce an electric signal with the same pattern as the electromagnetic wave. This is how radio signals and other communications signals are transmitted and received. For maximum sensitivity, the size of the antenna should be of the same order as the wavelength of the waves.

Infrared radiation

Infrared (IR) radiation can be detected with heat-sensitive devices such as thermocouple detectors and bolometers. A bolometer contains a blackened metal strip whose temperature rises when IR radiation falls on it. The temperature change is detected by measuring the change in electrical resistance of the strip. IR radiation can also be detected with semiconductor devices such as photovoltaic cells and photoconductors. When IR radiation falls on the surface of a semiconductor its energy may produce an emf and/or decrease the electrical resistance of the material. Devices based on these principles can be used to construct night vision cameras, which create images from the heat emitted by objects such as the human body.

Photographic films sensitive to IR radiation can be used to take IR pictures that reveal 'hot spots' in the landscape.

Visible light

Photographic films detect light by the chemical changes it produces in light-sensitive chemicals, such as silver halides. Light is also detected by the photoelectric effect (see Chapter 9) in which its energy causes electrons to be emitted from metal surfaces. In a photomultiplier tube these electrons are collected and the current amplified to produce an electric signal. Semiconductors are used to produce photovoltaic cells, which generate a current when light falls on them, and photoresistors in which incident light causes a change in electrical resistance.

Ultraviolet

Ultraviolet (UV) radiation can be detected in similar ways to visible light by using photographic films and appropriately designed electronics devices. UV radiation may also be converted to visible light by **fluorescence** before detection. A fluorescent material is one that absorbs the energy of UV light then re-emits it as visible light.

The inner surface of a fluorescent tube is coated with a fluorescent material. The tube is filled with a gas that emits UV light when made to conduct by a high voltage. The UV light causes the coating to emit visible light.

X-rays and gamma rays

In hospitals, the X-rays used to observe broken bones are detected by their action on

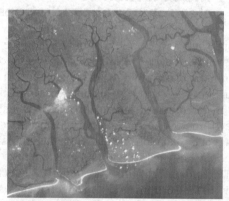

Fig. 4.3 An infrared photograph of a landscape

Fig. 4.4 Fluorescent light tube

Table 4.1 The detection of electromagnetic radiation

Radiation band(s)	Detection method
Radio and microwave	Antennae and aerials
Infrared	Heat sensors (thermocouple detectors and bolometers), photography, semiconductor devices
Visible	Photography, photoelectric effect, semiconductor devices
Ultraviolet	Photography, photoelectric effect, semiconductor devices, fluorescence
X-rays	Photographic emulsions, ionisation detectors
Gamma	Ionisation detectors, scintillation counters

specially designed photographic emulsions. This high energy radiation may also be detected by its ability to ionise gas atoms producing a pulse of electric current in a gas placed between two electrodes.

The Geiger counter uses this principle to detect the presence of both X-rays and gamma rays. The highest energy gamma rays produce scintillations – pulses of visible light – when they are incident on certain materials. By counting the rate of charge pulses or voltage pulses, or by measuring the scintillations of the light emitted, the number and energy of gamma ray photons striking an ionisation detector or scintillation counter can be found.

Table 4.1 gives a summary of the different methods needed to detect different electromagnetic radiations.

4.4 Applications of electromagnetic radiations

Gamma rays

Gamma rays are produced by energy changes in the nuclei of radioactive atoms. They carry a great deal of energy and can pass through thick layers of matter. They have harmful effects on living cells – they are used to sterilise food and medical instruments, kill weevils in grain and make pest insects sterile before releasing them into the wild. The damaging effects of gamma rays can be put to use to treat cancer. Beams of gamma radiation directed at a tumour can destroy cancer cells. Gamma ray sources are used in engineering to detect hidden cracks in structures (in a similar way to taking an X-ray) and to measure the thickness of materials by measuring how much radiation is transmitted.

X-rays

X-rays are produced in the laboratory by bombarding a metal with high-energy electrons inside an X-ray tube. Like gamma rays, X-rays are very penetrating. Because they are absorbed more strongly by dense materials, such as bone, they can be used to take pictures of the inside of our bodies and other objects which light rays cannot pass through. The denser regions appear as shadows on an X-ray film placed on the opposite side of the object from the X-ray source. The wavelengths of X-rays are approximately the same as the dimensions of atoms and molecules. This means that X-rays are diffracted as they pass by atoms in solids. Scientists can use the patterns of interference maxima produced when a beam of X-rays is passed through a crystal to measure the spacing of the atoms and to deduce the arrangement of atoms in molecules. This branch of science is called X-ray crystallography.

Ultraviolet rays

Ultraviolet radiation can be produced by passing an electric current through a low-pressure gas that contains heavy-metal atoms such as mercury. Electrons in the mercury atoms gain energy from the current, then emit it again as UV light. We cannot detect UV light with our eyes, but when it falls on some materials it makes them fluoresce – the UV radiation is absorbed and its energy re-emitted as visible light. The ink from the 'invisible' pens used to make security markings on valuables fluoresces under a UV lamp. The Sun is a natural source of UV radiation. The Earth's atmosphere filters out much of this radiation, but there is still sufficient in bright sunlight to damage your skin. Too much exposure to UV radiation can be harmful. It causes sunburn and, in large doses, skin cancer. Insects' eyes can detect UV light that is invisible to human eyes. Flying insects are attracted to UV lamps.

Visible light

Visible light makes up only a tiny part of the electromagnetic spectrum but it is crucial to almost all living creatures. Visible light provides the energy for plant growth. Most living things respond to light and many use it as their main way of sensing the world. We use light to create and capture images with film and video cameras, and to gather information about the stars using giant telescopes. We perceive different frequencies of light as different colours but it is impossible to describe what this means – try to imagine how you would explain the idea of 'red' or 'yellow' to someone who has always been blind.

Infrared rays

Infrared rays form a band just beyond the red end of the visible region of electromagnetic spectrum. We perceive them as heat when they are absorbed by our skin, or by other materials. When we sit by a glowing fire we are warmed mainly by IR radiation. Imaging devices can be made which convert IR images into visible images. These can be used to seek out 'hot spots' in our bodies for medical diagnosis, to take pictures of the Earth from space to show variations in vegetation, or as night sights by wildlife photographers, the police and the armed forces.

The greenhouse effect

Glasshouses (greenhouses) are used to grow plants in cool climates. Shorter wavelength infrared rays and light rays from the Sun can pass through the glass or plastic panels of a greenhouse and are absorbed by the soil and plants, raising their temperature. The soil and plants in turn emit longer wavelength infrared rays which cannot penetrate glass. The inside of the glasshouse stays hot. (Very hot objects, like the Sun, emit short wavelength infrared rays; cooler objects, like soil, emit longer wavelength infrared rays.)

Carbon dioxide and other gases in the atmosphere help the atmosphere to act in a similar way to a greenhouse. The atmosphere lets infrared and light rays pass through to the surface of the Earth. There they are absorbed and converted to internal energy of the plants and land mass. The warmer land now radiates longer wavelength infrared radiation. This is absorbed by carbon dioxide and water vapour in the atmosphere. These gases are warmed up as a result and they in turn radiate energy in all directions as infrared rays. Some of this energy is radiated back to the Earth. Without the carbon dioxide and water vapour this energy would have been lost – so the Earth is warmer as a result.

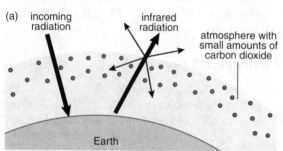 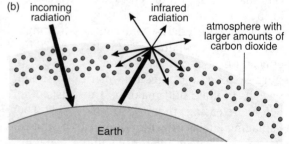

Fig. 4.5 The greenhouse effect
(a) Not much infrared radiation is absorbed
(b) Much more infrared radiation is absorbed and some of it is scattered back to earth

As a result of the burning of fossil fuels, the amount of carbon dioxide in the atmosphere is slowly increasing. It is estimated that burning fuels release about 18 000 million tonnes of carbon dioxide into the air each year. Many scientists think that the increasing amount of carbon dioxide is responsible for the rising temperature of the atmosphere and sea. Such changes are likely to alter the Earth's climate and raise the level of the sea. These processes are described as 'global warming' and 'climate change'.

> **Activity 4.2 Investigating the greenhouse effect**
>
> Use the library and/or the Internet to undertake further research into the greenhouse effect and global warming. Prepare a poster for display in the school hall on this topic. Your poster should provide information on:
> - the physics of the greenhouse effect;
> - the evidence for global warming;
> - the causes and consequences of global warming;
> - the steps that individuals and governments should take to tackle the problem.

Microwaves

Microwaves are produced by electrons oscillating at very high frequencies, for example inside electronic components called klystrons and magnetrons. They are used to cook food, send signals via mobile phones and to operate radar systems. The microwaves in a microwave cooker are generated at the same frequency as a natural vibration frequency of water molecules. Water molecules in food absorb the microwave energy through resonance and the food becomes hot. Radar is an echolocation system, similar to sonar, but uses microwaves instead of sound. The radar antenna emits pulses of microwave radiation as it turns. Echoes of the pulses are used to build up a picture showing the movements of aircraft, ships or rain clouds.

Radio waves

Radio waves are generated by antennae fed with high-frequency alternating currents from powerful transmitters. A radio wave incident on a receiving antenna induces a signal in the antenna with the same frequency. This signal is detected by resonance with a suitable electrical circuit in the receiver. The wavelengths used to transmit radio broadcasts are divided into a number of 'wavebands'. The long, medium and short wavebands are reflected from the upper layers of the Earth's atmosphere. This enables the reception of broadcasts in these bands from hundreds, or even thousands, of miles away. The shorter wavelengths used to transmit television and FM (frequency modulated) radio waves are not reflected by the atmosphere and so the receiver must not be shielded from the source, for example by high ground.

4.5 Solving problems involving $c = f\lambda$

> **Example**
>
> Orange light has a frequency of 6.0×10^{14} Hz. Calculate the wavelength of this light, given that the speed of light in air is 3×10^8 m s^{-1}.
>
> **Answer**
> $c = f\lambda$, so
> $$\lambda = \frac{c}{f} = \frac{3 \times 10^8}{6 \times 10^{14}} = 5 \times 10^{-7} \text{ m}$$

> **Exercise 4.1**
>
> 1. Identify the regions of the electromagnetic spectrum to which radiations with the following wavelengths belong. *(i)* Calculate the frequency of each radiation. *(ii)* Explain how it could be detected. *(iii)* What application might radiation of this wavelength have?
>   (a) $\lambda = 1$ m
>   (b) $\lambda = 10^{-2}$ m
>   (c) $\lambda = 10^{-5}$ m
>   (d) $\lambda = 10^{-10}$ m
> 2. Radio stations broadcast on the following frequencies:
>   (a) 2×10^5 Hz (200 kHz)
>   (b) 5×10^6 Hz (5 MHz)
>   (c) 3×10^8 Hz (300 MHz)
>   Calculate the wavelength in each case.

Summary

- An electromagnetic wave consists of vibrating electric and magnetic fields at right angles to each other – they carry energy as they travel through space.
- All electromagnetic radiation travels with the same speed through a vacuum, the speed of light.
- The speed of light, c, in a vacuum is 3×10^8 m s^{-1}.
- For electromagnetic waves travelling in a vacuum, $c = f\lambda$.
- Long wavelength electromagnetic waves have low frequencies; shorter wavelengths have higher frequencies.
- The energy carried by electromagnetic waves depends on their frequency. The radiations at the high frequency/short wavelength end of the spectrum (gamma and X-rays for example) carry far more energy than radiations at the lower frequency/longer wavelength end of the spectrum (radio waves for example).
- The electromagnetic spectrum consists of all possible wavelengths of electromagnetic radiation. Gamma rays are the shortest-wavelength, highest-energy form of electromagnetic waves. The other bands of the electromagnetic spectrum are, in increasing wavelength, X-rays, ultraviolet rays, visible light, infrared radiation, microwaves and radio waves.

Review questions

1. **(a)** State the properties of electromagnetic waves.
   (b) Draw a chart of the electromagnetic spectrum, labelling its components and giving their frequency and wavelength ranges.
   (c) Describe four applications of electromagnetic radiation from different regions of the spectrum.
   (d) Describe three methods used to detect electromagnetic radiation from different regions of the spectrum.

2. **(a)** State the relationship between wavelength, frequency and speed of electromagnetic waves.
   (b) What is the wavelength, in metres, of soft X-rays of frequency 2×10^{17} Hz?
   (c) What is the wavelength of green light of frequency 5.6×10^{14} Hz?
   ($c = 3 \times 10^8$ m s^{-1})

3. In a thunderstorm, we see a flash of lightning before hearing the accompanying thunder.
   (a) Explain this in terms of various wave speeds.
   (b) Describe how this observation can be used to estimate the distance of an observer from a storm.

4. **(a)** Radio waves often bend around obstacles while light waves are usually observed to travel in a straight line. If both waves are electromagnetic, why the difference?
   (b) White light is viewed through *(i)* a prism and *(ii)* a diffraction grating. Explain what is seen in each case.

5 Electromagnetic induction

Learning objectives

After completing the work in this chapter you will be able to:

1. perform and describe simple experiments to illustrate electromagnetic induction
2. state the factors affecting the magnitude and the direction of the induced emf
3. state the laws of electromagnetic induction
4. describe simple experiments to illustrate mutual induction
5. explain the working of an alternating current (a.c.) generator and a direct current (d.c.) generator
6. explain the working of a transformer
7. explain the applications of electromagnetic induction
8. solve numerical problems involving transformers.

5.1 Investigating electromagnetic induction

In Form Two you learned that an electric current produces a magnetic field. At the beginning of the nineteenth century, the scientist Michael Faraday posed the following question: if electricity produces magnetism – can magnetism be used to produce electricity? Through careful experiments Faraday discovered how a current can be induced (generated) by changing the magnetic field through a coil, or by moving a conductor through a magnetic field. This process is called **electromagnetic induction**. Faraday's discoveries were the starting point for the modern technological age – today we use electricity produced by electromagnetic generators to power everything from simple domestic light bulbs and heaters, to complex computers and huge manufacturing machines.

Activity 5.1 Experiments to illustrate electromagnetic induction

You will need:
- a sensitive galvanometer (milliammeter)
- varnished copper wire
- bar magnets
- plastic tape
- a sharp knife
- laboratory clamps.

Experiment 1
1. Wind 10-turn and 20-turn coils of copper wire as shown in Fig. 5.1. (Hint: use a broom handle or similar rod to help wind each coil.)
2. Use the sharp knife to scrape the varnish from the wire at both ends of the coils so that good electrical connections can be made.

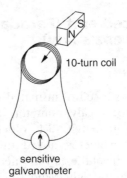

Fig. 5.1 Investigating induction in a coil

3. Connect the 10-turn coil to the galvanometer.

(continued)

Activity 5.1 (continued)

4. Plunge one pole of the magnet into the centre of the coil, hold it stationary, and then remove it again. Observe the size and direction of the galvanometer deflection. How is the galvanometer deflected when:
   - the pole is moving into the coil?
   - the pole is stationary?
   - the pole is moving out of the coil?
5. Investigate the effect of the following on the size and direction of the deflection obtained:
   - the speed at which the magnet is moved,
   - using the opposite magnetic pole,
   - increasing the number of turns in the coil (replace the 10-turn coil with the 20-turn coil),
   - keeping the magnet stationary and moving the coil.
6. Write notes on your observations.

Experiment 2

1. Clamp two magnets with opposite poles facing each other to produce a strong magnetic field as shown in Fig. 5.2.
2. Connect a length of wire to the galvanometer and place the wire in the field.
3. Move the wire rapidly out of the field, and then back in again. Observe the size and direction of the galvanometer deflection. How is the galvanometer deflected when:

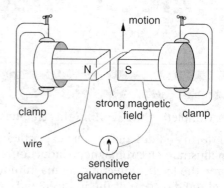

Fig. 5.2 Investigating induction in a wire

   - the wire is stationary?
   - the wire is moving out of the field?
   - the wire is moving into the field?

How does:
   - the size of the deflection depend on the speed at which the wire is moved?
   - the size of the deflection depend on the strength of the magnetic field? (You can move the poles apart to decrease the field strength between them)

4. Write notes on your observations.

Discussion

You have just repeated experiments first performed by Faraday and others. The significance of your findings is discussed in Section 5.2.

5.2 Induced emf (Faraday's law and Lenz's law)

Induced emf

Fig. 5.3 shows a coil of insulated copper wire connected to a galvanometer, as in your investigation in Activity 5.1. When the north pole of a bar magnet is moved into the coil of wire, the needle of the galvanometer is seen to be deflected to the right. When the bar magnet is stationary, however, the needle returns to its zero position. When the bar magnet is removed from the coil the needle is deflected to the left. Thus, whenever the magnet moves a current is produced in the coil. In other

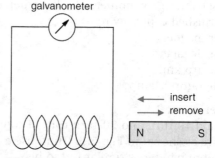

Fig. 5.3 A magnet moving inside a coil induces an emf in the coil

words, an emf (**electromotive force**) is induced in the coil. A similar result can be obtained by moving a coil of wire over a stationary magnet.

When Faraday first did this experiment he discovered that the strength of the emf, and hence the size of induced current, was increased by:
- moving the coil or magnet faster;
- increasing the strength of the magnet;
- increasing the number of turns of wire in the coil.

Faraday had discovered that electromagnetic induction occurs only when the magnetic field through a circuit is *changing* – either by increasing or by decreasing in strength. The more rapidly the field changes, the greater the emf induced in the coil.

Magnetic flux

In Form Two you saw that the strength of a magnetic field can be indicated by using lines of force. The field is strongest where the lines of force are closest together (where their density is greatest). Fig. 5.4 shows how the number of lines of force passing through a coil placed near a bar magnet depends on its position. Faraday described the number of lines of force passing through a loop of a given area as the **magnetic flux** through that area. The flux may be increased by:
- moving the coil to a location where the flux density (the number of lines of force per unit area) is greater;
- increasing the area of the coil;
- increasing the number of turns in the coil.

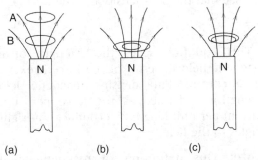

Fig. 5.4 The flux linked with the coil (the number of force lines that pass through it) may be altered by changing its (a) position, (b) area or (c) number of turns.

Faraday's law

In terms of magnetic field lines, Faraday's discoveries may be summarised as follows – an electromotive force is induced whenever there is a change in the number of lines of force passing through (linked with) a coil, in other words when there is a change in the magnetic flux through the coil. The emf induced is proportional to the rate at which the flux changes.

An alternative picture, which also fits the experimental observations correctly, is that an emf is induced in a conductor as it 'cuts' through lines of force. For example, as the coil in Fig. 5.4(a) moves from position A to B, the flux through the loop increases from 1 to 3 lines. Two lines of force have cut through the coil as they 'move' from the outside to the inside. The greater the speed at which lines are cut in this way, the greater the emf.

These discoveries are summarised by Faraday's law which states that:

the induced emf is proportional to the rate of change of flux linkage, or the rate of cutting across lines of force.

Lenz's law

The experiments in Activity 5.1 showed that a current flowed in one direction when the pole of the magnet was plunged into the coil, and in the opposite direction when the same pole of the magnet was pulled out again. The law that gives the direction of the induced current was formulated by Heinrich Lenz in 1834. Lenz's law states that:

the induced current is in such a direction as to oppose the change producing it.

In Fig. 5.5(a) the north pole of a bar magnet is pushed towards a solenoid. Lenz's law tells us that the induced current must be such that it turns the coil into an electromagnet with a north pole at the end facing the north pole of the magnet. The magnet is repelled and its motion is thus opposed.

When the magnet is pulled out of the solenoid, as in Fig. 5.5(b), the induced current turns the end of the coil facing the north pole of the magnet into a south pole, which attracts the north pole and opposes its motion. The direction of the induced current is, therefore, opposite to that obtained when the magnet moves towards the solenoid.

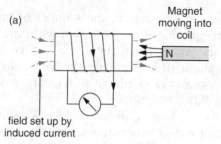
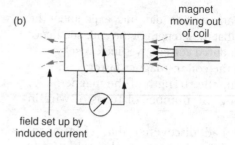

Fig. 5.5 Lenz's law
(a) Magnet repelled by the field of the solenoid
(b) Magnet attracted by the field of the solenoid

Emf in a moving wire

Experiment 2 in Activity 5.1 demonstrates that an emf is also induced in a wire when the wire moves in a magnetic field and cuts through lines of force. Fig. 5.6 shows a long copper wire connected to a galvanometer, G. Part of the wire, AB, is in a magnetic field. If AB moves at right angles across the magnetic field lines, a current is induced in the wire. A forward movement produces a current in the opposite direction to a backward movement. There is no current induced if the wire is stationary or moves parallel to the field so that it is not cutting field lines. The faster the movement of the wire, the greater is the emf produced.

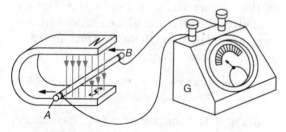

Fig. 5.6 Producing an emf by moving a wire in a magnetic field

Magnetic flux density (field strength)

The magnetic flux density is a measure of the strength of a magnetic field at a point. This is indicated by the density (the spacing) of the lines of force at that point. Where the density is high, the lines of force are closely spaced and the field strength is high.

In Form Two you learned that a magnetic field exerts a force on a wire carrying a current at right angles to the field. Experiments show that this force is proportional to the flux density (the strength of the field), the current and the length of the conductor. For a straight conductor of length l which is carrying a current i at right angles to the field:
$F = Bil$

Note: the direction of the force can be found using Fleming's left-hand rule.

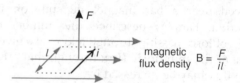

Fig. 5.7 The unit of magnetic flux density is defined in terms of the force on a current-carrying conductor in a magnetic field

B is the symbol for flux density (the strength of the magnetic field). From this equation we can see that the flux density is given by:

$$B = \frac{F}{il}$$

This equation gives the strength of a magnetic field in terms of the force it exerts on a current. Flux density (magnetic field strength) is defined as the force per unit current per unit length of a conductor at right angles to the field.

Note: This definition of magnetic field strength can be compared to the definitions of gravitational field strength and electric field strength which you have met previously:
- gravitational field strength, g, is the force per unit mass (unit N kg^{-1});
- electric field strength, E, is force per unit charge (unit N C^{-1})).

The SI unit of flux density is the tesla (T).

The flux density is one tesla when the force on a wire carrying a current of 1 ampere at right angles to the field is 1 newton per metre of length (1 T = 1 N A^{-1} m^{-1}).

Factors affecting the emf induced in a moving wire

In Activity 5.1 you found that the emf induced in a straight wire moving at right angles to a magnetic field increases with the strength of the field and the speed of the wire – it also increases with the length of the wire. These results are summarised by the situation shown in Fig. 5.8.

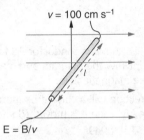

Fig. 5.8 Emf induced in a wire

A wire of length l moves at speed v at right angles through the field of strength B. Experiment shows that the emf, E, induced is given by:
$E = Blv$

This equation is consistent with Faraday's law, which states that the induced emf is proportional to the rate of cutting of lines of force, since:
- increasing B increases the number of lines of force to be cut;
- a longer wire cuts more lines of force than a short one;
- when the wire is moving faster it cuts lines of force more quickly.

An emf occurs in the wire whether or not the ends are connected in a complete circuit. When the ends of the wire are in a complete circuit, a current also flows. This can be shown by connecting the ends of the wire to a galvanometer (as in Activity 5.1).

Fleming's right-hand rule

Fleming's right-hand rule, illustrated in Fig. 5.9, gives the direction of the current induced by the motion of a conductor in a magnetic field. Be careful to distinguish between this and Fleming's left-hand rule, which gives the direction of the force on a current-carrying conductor in a magnetic field.

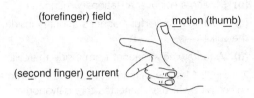

Fig. 5.9 Fleming's right-hand rule

The direction of the induced current can be found using Fleming's right-hand rule:
- the *F*irst finger points in the direction of the magnetic *F*ield;
- the thu*M*b points in the direction of the *M*otion;
- the se*C*ond finger shows the direction of the *C*urrent.

Note: Fleming's right-hand rule is used with generators (dynamos). Fleming's left-hand rule is used with motors.

Example

A wire 15 cm long is moved upwards at a speed of 100 cm s^{-1} at right angles to a magnetic field of strength 0.01 T. The field goes from left to right. Find the emf induced in the wire and illustrate its direction by a diagram.

Answer
Emf = Blv = 0.01 × 0.15 × 1 = 0.0015
    = 1.5 × 10^{-3} V.
Using Fleming's right-hand rule, we find that the induced current goes into the page along the wire, as shown in Fig. 5.10.

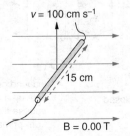

Fig. 5.10

Exercise 5.1

1. Which one of the following actions does not cause an induced emf to be set up in a coil of wire?

(a) Pushing a magnet into a stationary coil.

(b) Moving a coil over a stationary magnet.

(c) Holding a powerful magnet stationary inside the coil.

(d) Withdrawing a magnet from inside the coil.

2. Fig. 5.11 shows a magnet being pushed into a coil of wire which is connected to a galvanometer. Which of the following statements is/are correct?

(a) The induced current will flow from A to B through the coil.

(b) The induced current will flow from B to A through the coil.

(c) No induced current will flow.

(d) End B will become a north pole.

3. A magnet is used to induce a current in a coil of wire. List three things that could be done to increase the current produced.

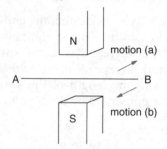

Fig. 5.12

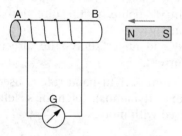

Fig. 5.11

4. (a) Fig. 5.12 shows a conductor AB in a magnetic field. Draw the diagram and mark the direction of the magnetic field.

(b) Which direction will a current be induced in the conductor AB when it is moved *(i)* into the page, *(ii)* out of the page?

5. A 5 cm length of wire passes through a magnetic field, at right angles to the field, with speed 12 m s^{-1}. The field strength is 1.5 T. Calculate the induced emf.

5.3 Mutual induction

Mutual induction is the induction of an emf in a coil by the changing magnetic field of a second coil placed nearby. Fig. 5.13 shows two coils placed side by side so that the magnetic field produced by one of the coils passes through the other. When a current flows through the first coil (the primary coil), it creates a magnetic field which passes through the secondary coil. If this field changes (by varying the current in the primary coil, or turning it on and off), an emf is induced in the secondary coil. It is the *change* in the field due to one coil which induces an emf in the other.

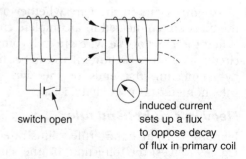

Fig. 5.13 Mutual induction

Activity 5.2 Investigating mutual induction

Your teacher will set up the demonstration shown Fig. 5.13.
- Observe the direction of the emf induced in the secondary coil when the current in the primary coil is turned on, and then off again.
- Record your observations.

According to Lenz's law, the induced current flows in such a direction as to oppose the change in the field that produces it. If the magnetic field becomes steady, no emf is induced. When an iron core is inserted through the two coils the mutual inductance is increased, because the field lines are concentrated in the core.

It is important to realise that mutual induction occurs only when the current in the primary coil is *changing*. An a.c.(alternating) current flowing in the primary coil is constantly changing and therefore always induces an a.c. current in the secondary coil; a d.c. current in the primary does not produce an emf in the secondary unless it changes.

Example

A primary coil AB and a secondary coil CD are arranged as shown in Fig. 5.14. The current in AB can be changed by sliding the rheostat R to the left or to the right.

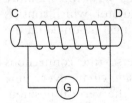
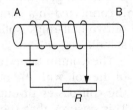

Fig. 5.14

(a) When the rheostat slides to the left, in which direction is the current in the secondary coil flowing?
(b) When the current through coil AB is steady, in which direction is the current in CD flowing? Explain your answer.
(c) Will coil CD be repelled or attracted to coil AB?

Answer

(a) The alphabet rule (Fig. 5.15) shows that end A of the primary coil is a south magnetic pole. As the rheostat slides to the left, the current increases and thus the field strength increases. Coil CD will produce a magnetic field to oppose this change. A current will flow in CD so that end D is a south pole. The induced current in CD will thus go from D to C.

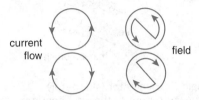

Fig. 5.15 The alphabet rule

(b) No current is induced in CD. To induce a current, the magnetic flux through the coil must be changing.
(c) CD will be repelled since ends A and D are both south poles.

5.4 Generators

A generator is a machine for converting mechanical (movement) energy into electrical energy. Generators make use of electromagnetic induction to produce an emf. A coil of wire is spun between the poles of a powerful magnet. The generator may be turned by petrol or diesel engines, by the energy of flowing water (hydroelectric power), by a steam turbine or, in the case of a bicycle dynamo, by muscle power.

A direct current (d.c.) generator produces a current that always flows in the same direction, like the current in a battery-powered circuit. An alternating current (a.c.) generator produces a current that changes direction constantly, travelling first one way and then the other – most large generators produce alternating current. The domestic electricity supply provides an alternating current which changes direction 50 times each second (50 Hz). Alternating current is preferred to direct current for electricity supplies because it is more easily transformed from one voltage to another.

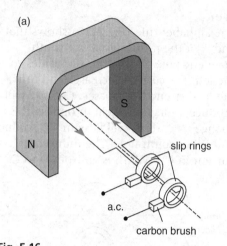
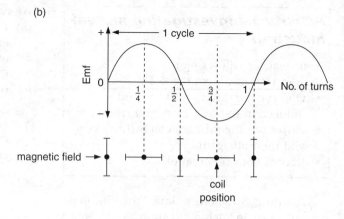

Fig. 5.16
(a) A simple a.c. generator (alternator)
(b) The current output of the a.c. generator

The a.c. generator

The principle of the a.c generator (also called an alternator) is shown in Fig. 5.16. In a practical generator, the coil would consist of many turns to increase the magnetic flux linkage – the coil is known as the armature winding. In Fig. 5.16(a) just one winding of the coil is shown for simplicity. The ends of the coil are fixed to two slip rings which rotate with the coil. The current passes to the outside circuit via carbon brushes which press against the slip rings.

As the coil rotates, the magnetic flux through it changes. The rate at which the flux changes is greatest when the coil is horizontal and is cutting the field lines most rapidly as shown in Fig. 5.16(a). The flux through the coil increases until the coil is in the vertical position. When the coil is vertical, its sides are moving parallel to the flux lines, not cutting them, so the induced emf drops to zero. As the coil passes the vertical, the flux through it starts to decrease again generating an emf in the opposite direction. The flux decreases more and rapidly until the coil is horizontal once more, and the emf reaches its maximum in the negative direction. Fig. 5.16(b) shows the output of the generator during one rotation of the coil.

As the coil continues to rotate, the emf and the current alternate at the rotation frequency. Increasing the speed of rotation increases the frequency of the a.c. generated.

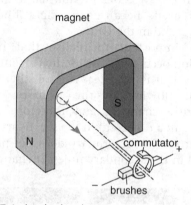

Fig. 5.17 A simple d.c. dynamo

The d.c. generator

In the d.c. generator (Fig. 5.17) the armature is connected to a commutator, which consists of two half-rings (split rings) mounted on the spindle.

The commutator reverses the connections of the coil with the outside circuit every time the coil passes through the vertical position. This ensures that the current in the outside circuit always flows in the same direction. You can see the close similarity between this and the simple motor you studied in Form Two. The current is maximum when the coil is horizontal and minimum when it is vertical. The induced emf, and hence the induced current, can be increased by:
- using a stronger magnet;
- increasing the number of turns in the coil;

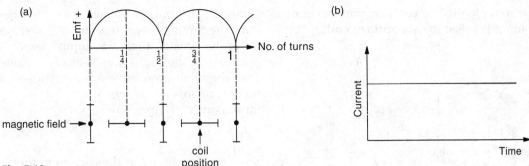

Fig. 5.18
(a) The current generated by a d.c. dynamo
(b) The current generated by a d.c. cell

- winding the coil on a soft-iron armature;
- rotating the coil at a higher speed.

Fig. 5.18 shows the output of the dynamo during one rotation of the coil and how it fluctuates compared to the smooth output of a d.c. cell.

5.5 Transformers

A transformer is a device that makes use of mutual induction to change voltage. It consists of two coils wound on an iron core (Fig. 5.20). The coil connected to the a.c. input is called

Activity 5.3 *Investigating a bicycle dynamo*

You will need:
- a bicycle with a dynamo
- a sensitive voltmeter
- a cathode ray oscilloscope if available.

Method
1. Turn the bicycle upside down and connect the output of the dynamo to the voltmeter.
2. Observe the output voltage as the dynamo is turned slowly.
3. Your teacher will connect the dynamo output to the CRO to show how the voltage varies during each revolution.

Exercise 5.2

1. Fig. 5.19 shows a diagram of a bicycle dynamo. Study the diagram and answer the following questions.

(a) What turns the driving wheel of the dynamo?

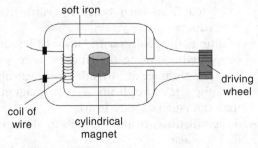

Fig. 5.19 A bicycle dynamo

(b) What is connected to the output of the dynamo?

(c) Briefly explain how the dynamo produces current.

(d) How could the output of the dynamo be increased?

2. (a) Draw a sketch graph to show how the emf of a simple a.c. generator varies with time over two full revolutions. Relate the positions of the coil to the values shown on your graph.

(b) Draw a second sketch graph showing what you would expect if the speed of rotation of the coil were doubled.

3. (a) Describe the main difference in construction between a d.c. generator and an a.c. generator.

(b) Sketch a graph to show how the current from a d.c. generator varies with time. How would the output change if a coil with twice as many turns were used?

the **primary coil**. The coil that provides the a.c. output is called the **secondary coil**.

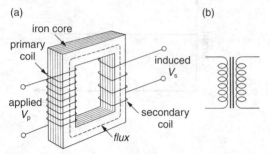

Fig. 5.20
(a) The structure of a transformer
(b) The circuit symbol for a transformer

A transformer can only operate on a *varying* voltage. A constant d.c. voltage connected to the primary coil will not produce any change in the magnetic field – so no voltage is induced in the secondary coil with d.c.

The effect of the iron core is to concentrate the flux produced by the primary coil so that virtually all of it passes through the secondary coil as well. As the primary flux is alternating, an alternating emf is generated in the secondary coil. Application of Faraday's law shows that, for an ideal transformer with no energy losses and in which all the flux of the primary passes through the secondary:

$$\frac{V_p}{N_p} = \frac{V_s}{N_s}$$

where V_p is the primary voltage, V_s the secondary voltage, N_p the number of turns of wire on the primary coil, and N_s the number of turns of wire on the secondary coil.

Example

A 200 V a.c. supply is connected to the primary coil of a transformer. The primary coil has 1000 turns. If the secondary coil has 2500 turns what is the output voltage?

Answer

$$\frac{V_p}{N_p} = \frac{V_s}{N_s}$$

$$\frac{200}{1000} = \frac{V_s}{2500}$$

Therefore, $V_s = \frac{200 \times 2500}{1000} = 500$ V.

Although the output voltage may be larger than the input voltage, the output *power* can never be greater than the input *power*. In practice, the output power is always smaller than the input power due to energy losses. If there is no loss in energy:

input power = output power

$$V_p I_p = V_s I_s$$

So, $\frac{I_p}{I_s} = \frac{V_s}{V_p}$, where I_p is the current in the primary coil and I_s the current in the secondary coil. This result can be applied to well-designed transformers in which the energy loss is small.

Energy loss in transformers

Most commercial transformers have an energy efficiency of about 90%. The sources of energy loss in a transformer are as follows.
- When an alternating current passes through the primary coil of a transformer, circulating currents, called eddy currents, are induced in the iron core – part of the input energy is lost as heat energy generated by these eddy currents.
- Heat is produced in the primary and secondary coils because they have resistance, however small.
- Some energy is required to repeatedly magnetise and demagnetise the core material – the energy lost in this way is called hysteresis loss.
- Some energy is lost if all the flux from the primary coil does not pass through the secondary coil – this is called flux leakage.

These energy losses may be minimised by:
- laminating the transformer core to reduce eddy currents – instead of a solid piece of iron the core is made from thin iron sheets insulated from each other by varnish or thin paper;
- winding the coils with low resistance copper wire – using the thickest wire and the least number of turns that is practical;
- making the core from a soft magnetic material (such as pure iron);
- designing the core and the coils to minimise flux leakage.

A transformer can be used to provide any a.c. voltage by constructing it with the appropriate ratio of N_p to N_s. If the secondary voltage is greater than the primary voltage, the transformer is called a **step-up transformer**. If the secondary voltage is smaller than the primary voltage, the transformer is a **step-down transformer**. Some transformers have a secondary coil with several leads connected to it, as shown in Fig. 5.21. These leads are called taps, or tappings, and make it possible to draw several different a.c. voltages from the same transformer.

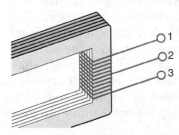

Fig. 5.21 A transformer with a centre tapping in the secondary coil

A transformer may have more than one secondary winding on the same iron core. If the windings are properly chosen, one secondary may act as a step-up transformer and the other as a step-down.

Example

If the voltage across terminals 1 and 3 in Fig. 5.21 is 9 V, what is the voltage across:
(a) terminals 1 and 2,
(b) terminals 2 and 3?

Answer

(a) This is a centre-tapped transformer so there are half as many windings between terminals 1 and 2 as there are between terminals 1 and 3. So the voltage will be 4.5 V.
(b) There is the same number of windings between terminals 2 and 3 as there are between terminals 1 and 2 – so the voltage will be the same, 4.5 V.

5.6 Applications of electromagnetic induction

The induction coil

Consider a transformer with a primary coil made of only a few turns and a secondary made with very many turns, as shown in Fig. 5.22.

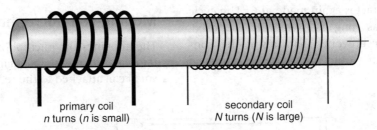

Fig. 5.22 A step-up transformer

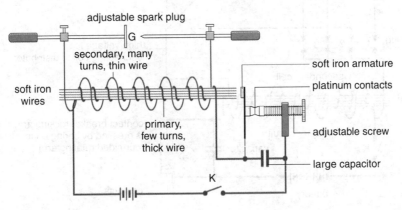

Fig. 5.23 An induction coil

Chapter 5: Electromagnetic induction 61

When a steady d.c. current flows in the primary coil, its magnetic field is constant and so no emf is induced in the secondary coil. But when the current in the primary coil is switched off or on, the magnetic field changes and, for a short time, an emf is induced in the secondary coil. Because the secondary coil has many more turns than the primary, the emf induced is much larger than the voltage applied to the primary. The ratio of the voltages is equal to the ratio of the number of turns in the coils.

The high voltage produced can be used, for example, to produce a spark. Fig. 5.23 shows an induction coil with the two coils wound one on top of the other.

When switch K is closed a current flows in the primary coil. The iron core acts as an electromagnet and attracts the armature, which moves towards the core and contact C opens. The current in the primary coil falls to zero. A large emf is induced in the secondary coil as the magnetic field increases and then decreases again. If the gap G is not too large then a spark is seen – to cause a one-centimetre spark in dry air needs about 30 000 V. A spring now pushes the armature back and contact is again made at C. The cycle is repeated as long as switch K is closed. The capacitor prevents sparking at the contacts C, which would otherwise be damaged quickly.

The ignition coil

An induction coil is used in petrol engines to make the sparks which ignite the mixture in the cylinders – Fig. 5.24 shows such a coil. Instead of being opened by the magnetic action of the core, the contacts are opened mechanically by a rotating cam. This is square-shaped for a four cylinder engine and hexagonal for a six-cylinder engine.

The cam is arranged so that a spark is produced in the correct cylinder at the correct time – just before the piston reaches the top of its compression stroke. The spark is made across the gap in a spark plug which has a central insulated electrode and a second electrode fixed to the metal body of the plug. A spark plug is screwed into the top of each cylinder.

For the engine to operate efficiently a good spark must be produced at the right moment. For this to happen, the contacts must be kept in good condition – the capacitor connected across them does not completely protect them from being damaged. The gap between the electrodes in the spark plug must also be kept at the right size.

Moving-coil microphone

The moving-coil microphone (Fig. 5.25) contains a thin metal foil diaphragm. There is a small coil attached to the rear of the diaphragm.

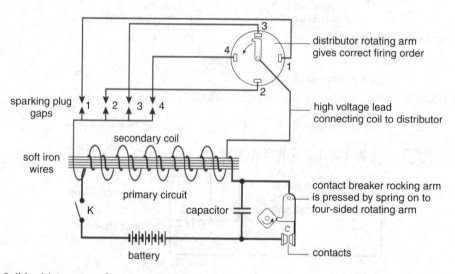

Fig. 5.24 Coil ignition system for a petrol engine

This coil is situated in a magnetic field provided by a cylindrical permanent magnet. Sound waves cause the diaphragm and coil to vibrate. As the coil moves in the magnetic field, a current is induced in it. This varying current can be amplified and heard in a loudspeaker.

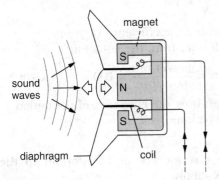

Fig. 5.25 A moving-coil microphone

(b) Both coils are stationary and S is turned on and off.
(c) With S closed, the variable resistance R is increased and decreased rapidly.
(d) S is left open and the resistance R is increased.

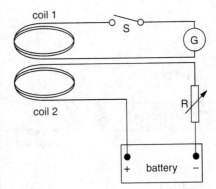

Fig. 5.26

5.7 Solving problems on transformers

Example

A step-down transformer has a primary coil of 1200 turns and a secondary coil of 40 turns. What is the voltage across the secondary coil when the primary coil is connected to a 200 V a.c. supply?

Answer

$$\frac{V_p}{N_p} = \frac{V_s}{N_s}$$

$$\frac{200}{1200} = \frac{V_s}{40}$$

$$V_s = \frac{200 \times 40}{1200} = 6.67 \text{ V}.$$

Exercise 5.3

1. The transformer described in the worked example above has a tapping on the secondary coil at 30 turns. What voltages could be obtained from the transformer? (Hint: voltages, not voltage.)

2. Two flat coils are mounted as shown in Fig. 5.26. Which of the following actions will cause the galvanometer to register a current?
(a) Coil 2 is stationary and coil 1 is moving upwards with S kept closed.

3. A transformer has a primary coil with 100 turns and a secondary coil with 250 turns. The primary voltage is 12 V.
(a) Is this a step-up or a step-down transformer?
(b) What voltage would be obtained from it?

4. A current of 2 A is passed through the primary coil (of 50 turns) of a transformer. The secondary coil has 400 turns. What current would be obtained from this transformer? State any assumptions you make.

5. A step-down transformer gives a current of 5 A at 12 V. If the primary voltage is 240 V, calculate:
(a) the primary current,
(b) the power input,
(c) the power output, assuming there are no power losses.

5.8 Project to construct a simple transformer

Fig. 5.27 shows two designs for simple transformers. The primary and secondary coils are wound with varnished copper wire (don't forget to scrape the varnish from the wire ends).

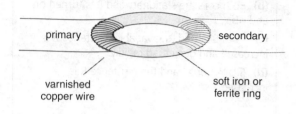

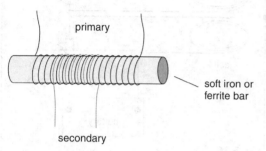

Fig. 5.27 Simple transformer designs

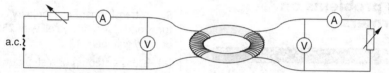

Fig. 5.28 Test circuit

Method

1. Construct one of these designs and test its performance using the circuit shown in Fig. 5.28.
2. Measure V_s and V_p. Does the equation
$$\frac{V_p}{N_p} = \frac{V_s}{N_s}$$
apply?
3. Determine the efficiency of your transformer from the equation:

$$\text{efficiency} = \frac{\text{power out}}{\text{power in}} \times 100 = \frac{V_s I_s}{V_p I_p} \times 100\%$$

4. Suggest how the efficiency of the design could be improved.

Safety

- Do not connect your transformer to the mains supply.
- Adjust the rheostat to restrict the current in the primary circuit to no more than 1 A.

Summary

- The phenomenon of inducing an emf in a conductor by a change in the magnetic field strength is called electromagnetic induction.
- When a conducting wire moves in a magnetic field, an emf is induced by electromagnetic induction.
- For a conductor moving at right angles to the magnetic field, emf = Blv.
- Fleming's right-hand rule applies to the generation of an emf.
- Faraday's law states that the induced emf is proportional to the rate of change of flux linkage, or the rate of cutting of lines of force.
- Lenz's law states that the induced current is in such a direction as to oppose the change producing it.
- The size of the induced emf can be increased by moving the conductor more quickly through the field, increasing the strength of the magnet or by increasing the number of turns in a coil.
- The phenomenon of inducing an emf in one circuit by a change of current in another circuit is called mutual induction.
- Transformers may be step-up or step-down, according to whether the output voltage is greater or smaller than the input voltage.
- The ratio of the input and output voltages is the same as the corresponding ratio of turns in the coils, $\frac{V_s}{V_p} = \frac{N_s}{N_p}$
- Energy loss in transformers arises from the heating effect of the current in the primary coil, eddy currents in the transformer core and magnetic hysteresis loss.

Review questions

1. The diagram below shows two identical coils of copper wire, P and Q, placed close to each. Coil Q is connected to a d.c. power supply and coil P is connected to a galvanometer G.

   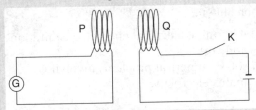

   (a) State and explain what would be observed on the galvanometer immediately switch K is closed, and then opened again.
   (b) How would the observations made in (a) differ if the number of turns in coil P were doubled? Explain your answer.

2. (a) State the laws of elecromagnetic induction.
   (b) The primary coil of a transformer has 2000 turns and carries a current of 3.0 A. If the secondary coil is designed to carry a maximum current of 30.0 A, calculate the minimum number of turns from which it may be wound.

3. The ratio primary turns: secondary turns for a transformer is 1:20. Assuming perfect efficiency, what currents flow in the primary and secondary windings if the primary voltage is 100 V and the resistance of the secondary circuit is 500 kΩ.

4. (a) List the sources of power loss which occur in a transformer and explain how these are minimised.
   (b) A transformer is connected to a 12.0 V, 30.0 W lamp from the 240 V mains. If the transformer is 75% efficient, determine the mains current.

5. The figure below shows a conductor in a magnetic field. It is moved downwards as shown. Show on a copy of the diagram the direction of the induced current in the conductor.

   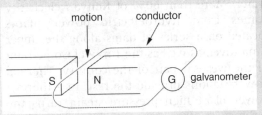

6 Mains electricity

Learning objectives

After completing the work in this chapter you will be able to:

1. state the sources of mains electricity
2. describe the transmission of electric power from the generating station
3. explain the domestic wiring system
4. define the kilowatt-hour
5. determine electrical energy consumption and cost
6. solve numerical problems involving mains electricity.

6.1 Sources of mains electricity

The mains electricity supply is generated by large electrical power stations. In a power station a.c. generators are turned by turbines. At a hydroelectric plant the energy to turn the turbines is provide by the pressure of moving water. More than 60% of Kenya's electricity comes from hydroelectric power plants located on a series of dams along the upper Tana River, as well as the Turkwel Gorge Dam in the western part of the country. Thermal power stations provide the rest. In a thermal power plant high pressure steam turns the turbines to drive the generators. The heat energy to boil water to produce steam may come from burning fossil fuels (oil, gas or coal), geothermal sources (hot underground rocks) or nuclear reactions. Successful geothermal power plants have been constructed at Olkaria. Their steam-powered generators

Fig. 6.1 Generators in a power plant

produce more than 10% of Kenya's electricity.

The waveform of the a.c. generated is a sine curve (see Fig. 6.2). The frequency of the a.c. supplied commercially in most of the world is 50 Hz. An alternating current or voltage is

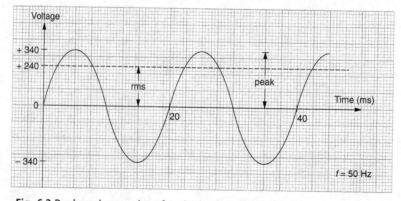

Fig. 6.2 Peak and rms values for the household mains supply

usually measured by its effective or **root mean square** (rms) value. This is the value which is equivalent to a steady direct current that gives the same heating effect. Suppose that a lamp is lit by a.c. and its brightness noted. If 0.3 A of d.c. produces the same brightness, the rms value of the a.c. is 0.3 A. Calculation shows that the rms value of a sine wave is related to the its peak value (the amplitude of the wave) by the equation

r.m.s. value = $\dfrac{\text{peak value}}{\sqrt{2}}$ = 0.707 × peak value

The rms value of the household mains supply is 240 V – the peak value is thus 240 ÷ 0.707 = 340 V. This means that mains wiring must be insulated to withstand at least 340 V. Meters that read a.c. voltages give rms values.

Note: Rms values are always used for calculations involving a.c. power, current and voltage.

Activity 6.1 *Visiting a local power station*

Research the location of your nearest power plant.
- What type of plant is it?
- Observe how the generators are operated and how the electricity generated is distributed from the power plant to the community.

6.2 Power transmission

Electrical power is transmitted from power stations to factories and homes by power lines. Alternating current is easily stepped up and down using transformers, so power transmission is almost always done using a.c.

Because of the enormous distance the current has to travel, the resistance of a power line cannot be neglected. When a current passes through a line, power is lost because of the heating effect of the current. The power loss is proportional to the square of the current and the resistance of the line – that is $P = I^2R$.

For the same amount of power delivered by a power line, the current is smaller if the voltage is higher. For this reason high voltages are used in power lines to reduce power loss. This is illustrated in Table 6.1, which shows the current needed to deliver power at a rate of 300 GW, and the power losses involved per kilometre of cable, for a range of voltages.

Table 6.1 Electrical power loss per kilometre for a range of transmission voltages

Power (GW)	Voltage (kV)	Current (A)	Power loss per km (kW)
300	66	4545	11 476
300	220	1363	1032
300	500	600	200

Of course, such high voltages raise installation costs, require expensive transformers and create problems of insulation along the cables. Nonetheless it is now common to transmit power at 500 kV or more.

Fig. 6.3 shows an experiment to demonstrate the power loss when power is transmitted at low voltage over long distances. The

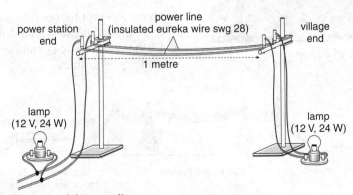

Fig. 6.3 A d.c. low-voltage model power line

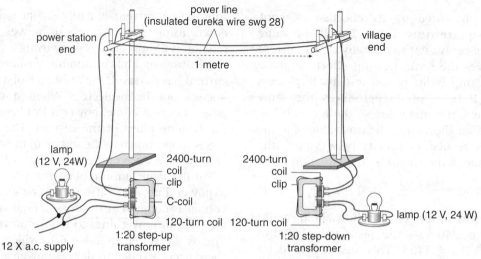

Fig. 6.4 A model a.c. power line using step-up and step-down transformers

long-distance transmission lines are simulated by using resistance wire. The light bulb at the village end is much dimmer than the one at the power station.

Fig. 6.4 shows a model a.c. transmission line. The voltage is raised by a step-up transformer at the power station. It is lowered by a step-down transformer at the village end. The lamp at the consumer end is now brighter than in the power line in Fig. 6.3. In stepping up the voltage, the transformer steps down the *current* and so the *power loss* is much smaller.

These experiments show why a.c. rather than d.c. is used for power transmission over long distances:

- a.c. voltages can be changed easily and without much loss of power by transformers;
- a.c. generators are also simpler, cheaper, more reliable and more efficient than d.c. generators.

The very high voltages associated with power transmission lines create a lethal hazard (Fig. 6.5).

- Never climb pylons.
- Do not fly kites near power lines.
- Do not fish with a rod and line under power lines.
- Never attempt to touch or retrieve an object from a power line with a pole or rope.
- Do not carry long ladders or similar objects under power lines.

Fig. 6.5 The high voltage of a power line is extremely dangerous

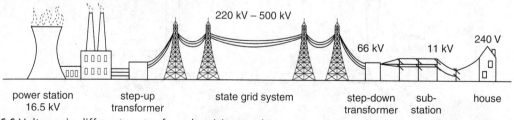

Fig. 6.6 Voltages in different parts of an electricity supply system

For safety purposes, electricity received by consumers should be at a lower voltage than that used for transmission. Because of this need to change voltages, electrical power is transmitted by a.c. so that transformers can be used. Fig. 6.6 shows the different voltages used in different parts of the transmission system.

In practice, electricity is generated by a.c. generators at power stations at between 11 kV and 33 kV. It is stepped up to a higher voltage (between 220 and 500 kV) by transformers. It is then transmitted across country by a grid (a network of power lines) to the various towns. The grid is fed by many power stations (some large and some small) and supplies every consumer connected to it. If one power station cannot work at full capacity, the others can share the extra load. The power lines consist of cables supported by tower-like pylons. Subsequently, the voltage is stepped down in successive stages (66 kV, 11 kV) at substations by transformers. Electricity is often conveyed to consumers at 240 V – this voltage is much safer, but still dangerous to life.

Example 1

Cables of resistance 2 Ω supply 2 kW of power. Calculate the power loss in the cable if power is transmitted at **(a)** 200 V and **(b)** 2000 V.

Answer

(a) $P = VI$

$I = \dfrac{P}{V} = \dfrac{2000}{200} = 10\,A$

Power loss $= I^2R = 10 \times 10 \times 2 = 200\,W$

(b) $P = VI$

$I = \dfrac{P}{V} = \dfrac{2000}{2000} = 1\,A$

Power loss $= I^2R = 1 \times 1 \times 2 = 2\,W$

Notice the much smaller power loss at the higher voltage.

Example 2

Figure 6.7 shows an electrical transmission system. The resistance of the transmission wire on each side is 4 Ω (so total resistance of line = (4 + 4)Ω). The 6 V, 24 W lamp is operated at its correct rating. Assuming that there is no energy loss in the transformers, calculate:
(a) the current through the lamp,
(b) the current through the transmission wire,
(c) the power loss in the transmission wire.

Answer

(a) $P = IV$

$I = \dfrac{P}{V} = \dfrac{24}{6} = 4\,A$

(b) For the step-down transformer

$I_p = I_s \left(\dfrac{N_s}{N_p}\right) = 4 \times \dfrac{1}{20} = 0.2\,A$

(c) $P = I^2R = 0.2^2 \times (4 + 4) = 0.32\,W$

6.3 The domestic wiring system

The electrical energy available in a household circuit originates in a power station. The current is almost always alternating current. Household electricity in Kenya is supplied at 240 V a.c. with a frequency of 50 Hz.

Plug wiring

The electric cable used in household circuits consists of three insulated wires. Each wire has a specific purpose and colour. Fig. 6.8 shows these three wires in a plug. When a three-pin plug is inserted into a power point, each pin in the plug connects to the corresponding wire in the power point.

The live wire is coloured brown (or red). This carries the alternating current to the appliance. The potential of the live wire varies

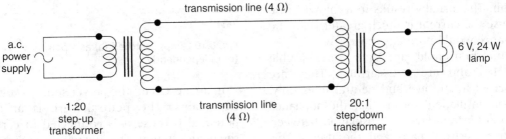

Fig. 6.7

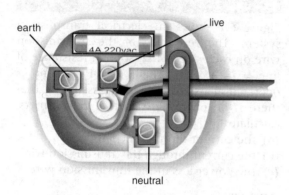

Fig. 6.8 A correctly wired plug

between positive and negative – therefore current flows to and fro through the circuit. If you touch a live wire you get a shock as the current passes through you to the earth. The shock can easily kill you.

The neutral wire is coloured blue (or black). It completes the circuit by providing the return path to the mains. The neutral wire is earthed at the local electricity supply substation, so its potential is 0 V. Although it carries alternating current when it is connected to the live wire through an appliance, there is no danger of electric shock if it is touched since it is at the same potential as a person who stands on the floor.

The earth wire is coloured green and yellow (or green). This wire is for safety purposes. In household circuits earth wires are connected to a common point in the ground underneath the house through a copper rod or water pipe.

Many electrical appliances have metal cases. If the insulation of the live wire becomes damaged and the wire makes contact with the case, there is a danger of electric shock. But if the metal case is connected to the earth wire the current will flow to earth through the wire, instead of through the person. This usually results in a 'blown' fuse because the current is much greater than the appliance would normally draw.

Some household appliances are double insulated and are not earthed. They are connected to the live and neutral lines only. 'Double insulation' means that the appliance has two separate layers of insulation between the live electrical parts and the user – this usually means that the appliance is enclosed in a plastic instead of a metal case. Such appliances do not need to be earthed because the case cannot become live, as it does not conduct electricity. Appliances that are double insulated are marked with the symbol shown in Fig. 6.9.

Fig. 6.9 Symbol on an electrical appliance which has double insulation

Activity 6.2 **Wiring a plug**

You will need:
- a three-pin plug
- a length of three-core flex
- a screwdriver
- wire strippers or a sharp knife.

Method
Your teacher will demonstrate how to wire the plug correctly. Follow his or her instructions to wire your plug.

Switches

A switch placed anywhere in a series circuit will stop the current flow. In a parallel circuit, each branch can have its own switch, or there can be one switch to control the whole circuit. Fig. 6.10 shows a circuit with two lamps controlled (a) as a pair and (b) independently.

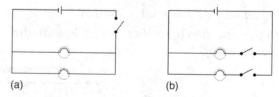

Fig. 6.10 Lamps switched (a) as a pair, (b) independently

In an a.c. circuit the power source is usually the mains. The neutral wire is at earth potential because it is connected to earth at the power station. The live wire (carrying a.c.)

is alternately above and below earth potential. Unless the switch is in the live wire the appliance itself will still be live when the current is switched off. This means that it is important to wire a domestic socket correctly. If the live and neutral wires are reversed then turning off the switch on the wall socket will not disconnect the appliance from the live terminal. It is therefore still possible to get a shock from it. This is illustrated in Fig. 6.11.

(a)

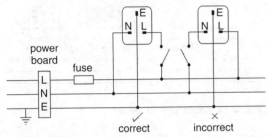

Fig. 6.11 Correct and incorrect switch locations

Fuses and circuit breakers

Damaged insulation or a loose connection can lead to a short circuit. A short circuit is an accidental connection between the live terminal and the neutral terminal, or the live terminal and earth, which bypasses the main circuits of an appliance. If the resistance of the short circuit is less than that of the appliance circuits, a large current flows through it.

When current flows through a resistor some energy is turned into heat. A large current flow, as the result of a short circuit, can make wires elsewhere in the circuit very hot and, if they are insulated, the insulation may begin to burn. Fires can start in this way.

To guard against fires from short circuits and other faults a **fuse** is put into the circuit. The simplest fuse is a piece of wire, or a thin strip, made of a metal with a low melting point. If the current rises too high the wire melts before the other wires can heat up dangerously.

A fuse is made so that it breaks the circuit at a specified current – for fuse wire this is often 5, 10, 15 or 30 A. These fuses are used in domestic circuits and can be replaced when they break. Various strip fuses are used in cars. Cartridge fuses, used in plugs, are rated at 3, 5 or 13 A – see Fig. 6.12(a). An alternative to a fuse is a **circuit breaker**, as shown in Fig. 6.12(b). This contains an electromagnetic

(b)
Fig. 6.12
(a) Cartridge fuses; (b) Circuit breakers

switch which operates automatically when the current exceeds a preset value. The circuit breaker has two advantages over a wire fuse – the current at which it trips can be set more precisely and once it has operated it can be reset, rather than replaced.

Household wiring

The wiring in a house is divided up into different circuits. There are separate circuits for lighting and for supplying power to other appliances through wall sockets. High-power devices, such as water heaters and cookers, are given individual circuits. All the different appliances we use in the house are designed to work at the same voltage, the voltage of the supply, which is 240 V. They must always be wired in parallel to ensure that the voltage delivered is correct.

Lighting circuit

Lights do not draw much current but, as with all domestic appliances, they must all be connected in parallel. In modern installations all light fittings must be connected to earth, but in older circuits this is not always the case. Fig. 6.13 shows a lighting circuit. All the switches are in the live line. A fuse is included in the circuit – this is often rated at 5 A. Such fuses are sometimes coded with a white dot.

Fig. 6.13 A lighting circuit

Power circuit

Appliances such as cookers, electric heaters, air-conditioning units, refrigerators, hair dryers and vacuum cleaners require larger currents than light bulbs. Some typical values are given in Table 6.2.

Table 6.2 Power ratings of domestic appliances

Appliance	Power rating (W)	Current (A)
Light bulb (typical)	100	0.42
Refrigerator	200	0.83
Power drill	500	2.1
Air conditioner	1200	5.0
Hair dryer	1200	5.0
Heater	2000	8.3
Vacuum cleaner	1350	5.6
Cooker	6500	27

Appliances like these must either be plugged into sockets or, in the case of cookers, wired into a separate circuit connecting directly to the supply. The total load on the supply at any one time can easily exceed 6000 W, even without cooker. If the voltage is 240 V this means that the current is:

$P = VI$

$I = \dfrac{P}{V} = \dfrac{6000}{240} = 25 \text{ A}$

To carry this current safely, the wires used for the main supply must be thicker than the wires in, say, a lighting circuit (so that their resistance is less). The power circuit for a cooker is protected by a 30 A fuse – these fuses are sometimes coded with a red dot.

Wall sockets are wired into power circuits in one of two ways – the spur system or the ring main. Fig. 6.14 shows a spur power circuit. In this circuit sockets are wired across the supply wires just as lights are wired. The fuse can be rated at 30 A.

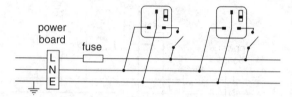

Fig. 6.14 A spur power circuit

Fig. 6.15 shows a ring main system. In a ring circuit both ends of the live, neutral and earth wires are connected to the power board to make complete loops. The sockets are then connected in parallel to the rings as shown in Fig. 6.16. The advantage of this is that current can flow through both paths to a socket. The wiring therefore does not have to be as heavy as, on average, each section of the wire carries less current.

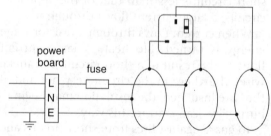

Fig. 6.15 A ring main

Appliances

Electrical appliances used in the home range from simple electric light bulbs, to heating and power devices such as kettles and hair dryers, to sophisticated electronic devices such as television sets.

Electric lamps

Incandescent light bulbs consist of a coil of fine resistance wire enclosed in glass bulb. The current flowing through the filament makes it

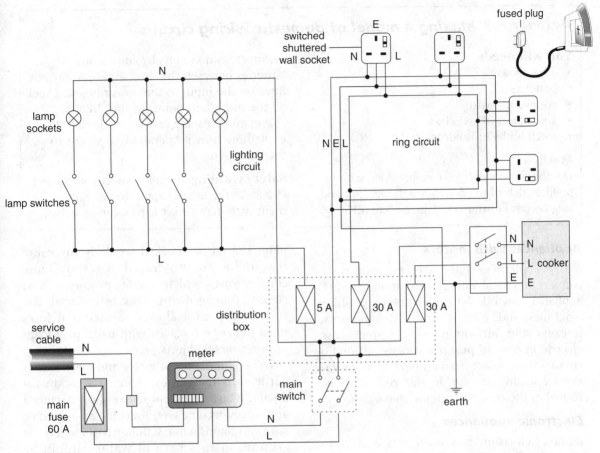

Fig. 6.16 A complete home wiring system

glow white hot. To prevent the filament being oxidised by the air, it is enclosed in a glass envelope filled with an inert gas. A glowing filament is shown in Fig. 6.17. Incandescent lamps have low efficiency as most of the energy is converted into heat rather than light.

Fluorescent lamps are more efficient. They produce light by passing an electric current through a gas containing mercury atoms trapped inside a glass tube. The current causes the mercury atoms to emit rays of ultraviolet light. The invisible ultraviolet rays strike a phosphor coating on the inside wall of the tube. The phosphor absorbs the ultraviolet light and re-emits the energy as visible light. A modern fluorescent lamp rated at 9 W gives out as much light energy as a 60 W incandescent bulb. Although they are more expensive to purchase initially, fluorescent lamps last longer than incandescent bulbs and, with their extra efficiency, are much cheaper to run overall.

Fig. 6.17 A glowing lamp filament

Cookers and heaters

These appliances rely on resistance heating in a similar way to the incandescent light bulb. Their heating elements glow red hot as a large current passes through them.

Chapter 6: Mains electricity 73

> **Activity 6.3** *Making a model of domestic wiring circuits*
>
> **You will need:**
> - a baseboard
> - batteries
> - connecting wire
> - a selection of switches
> - torch bulbs in holders.
>
> **Method**
> Use the components you are provided with to build models of the domestic lighting, spur and ring circuits illustrated in Figs. 6.13 to 6.15.
>
> Power your circuits by connecting batteries between the 'live' and 'neutral' lines on the input to the power board. Label all the important points in the circuits.
>
> Prepare a display card for each circuit to explaining how it operates to a visitor to the classroom.
>
> **Safety warning:** do not connect your model circuits to the mains supply. Never experiment with mains electricity, it can be lethal.

Appliances with motors

Many appliances have electric motors to convert electrical energy into motion. Such appliances include fans, hair dryers, washing machines and electric drills. Refrigerators, freezers and air-conditioning systems use electric motors to pump a coolant around a circuit. The coolant extracts heat by evaporation, releasing it outside the region to be cooled as the coolant is condensed again.

Electronic appliances

Radios, television sets, hi-fi sets and computers have complex *electronic* circuits that use electricity to store and process information in the form of electrical signals. These appliances generally do not draw as much power as heating or lighting appliances, but their operation is more likely to be affected by interruptions or surges in the power supply. Some of the components used to build electronic circuits are discussed in more detail in Chapter 11.

Domestic electric safety

Mains electricity is dangerous. Electric shocks, burns and fire can result from faulty circuits or appliances. To avoid accidents, wiring and plugs should be checked regularly to ensure that there are no frayed wires or loose connections, which might produce short circuits. Damaged wires must be replaced. The correct fuse must always be used and fuses must never be replaced with nails, paperclips or other metal objects.

The combination of water and electricity is particularly dangerous. Water is a good conductor, which readily produces short circuits if appliances become wet. Touching an appliance with wet hands while standing barefoot on the ground, or in a bath of water, should be avoided at all costs. The water provides an excellent connection between the appliance and your skin and a large current will flow through your body to earth. Skin resistance is lower if it is wet or broken. The effect of an electric current on the body depends on the resistance of the body and the floor, and the path the current takes. The lower the resistance is then the greater the current which will flow. Skin can be badly burned by electric current. Current flowing internally causes muscles to contract uncontrollably. The most dangerous path for the current to take is from one limb to another across the chest, since this has most effect on the heart – see Table 6.3.

> **Activity 6.4**
>
> Design an electrical safety poster for display in the classroom. Make your poster as visual as possible.
>
> Your poster should highlight the dangers of electricity and summarise the safety precautions outlined in this section.

To save a person who is in danger from electrocution, you should first disconnect the supply. If this is not possible use a dry stick to push the person off the wires. You *must not* touch the person before either the supply has been switched off or the person has been disconnected from it – if you do, the current might flow through you as well.

Table 6.3 The effects of electric current on the human body

Current (mA)	Effect on the body
1	Maximum safe current
2–5	Begins to the felt by most people
10	Muscular spasms, unable to let go, could be fatal
100	High chance of being fatal if through the heart
Any current greater than 25 mA is likely to be fatal to the human body	

6.4 The consumption and cost of electrical energy

Energy is measured in joules (J). Power is measured in watts (W). You will recall that power is the rate of energy transfer (power = energy ÷ time) and that 1 W is the power when 1 J of energy is being transferred in 1 s:
$1\text{ W} = 1\text{ J s}^{-1}$

It follows that the amount of energy consumed by a device (an electric lamp for example) is equal to its power multiplied by the time for which it has been operating:
energy = power × time

An alternative unit for energy is thus the watt-second (W s) since
$1\text{ J} = 1\text{ W} \times 1\text{ s}$

In 10 seconds a 100 watt light bulb consumes $100\text{ W} \times 10\text{ s} = 1000\text{ W s}$ of electricity.

But the watt-second is too small a unit for practical use. The unit used in practice is the **kilowatt-hour** (kW h). This is not an SI unit but it is used all over the world.

One kW h is the amount of electrical energy delivered in one hour (3600 s) at power of 1 kW (1000 W):
$1\text{ kW h} = 1000\text{ W} \times 3600\text{ s} = 3.6 \times 10^6\text{ J}$

If we know the power rating of an appliance and the time for which it is used, the consumption in kW h is easily calculated. Electricity companies charge by the kilowatt-hour for the use of the power they produce – they call 1 kW h a **unit**. So the cost of using an appliance for a certain length of time is worked out using

total cost = number of kW h × cost per unit.

An electricity meter (see Fig. 6.18) records the total number of units consumed on the premises where it is located.

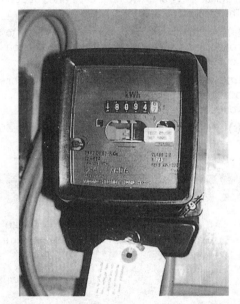

Fig. 6.18 An electricity meter reading 'units' (kW h)

6.5 Problems on mains electricity

Example

If one unit of electricity cost sh 3.5 calculate the cost of using **(a)** a 60 W light bulb **(b)** a 1 kW heater for 30 minutes each.

Answer

(a) Power = 60 W = $\dfrac{60}{1000}$ kW

Time = 30 min = $\dfrac{30}{60}$ h

Energy consumed = $\frac{60}{1000}$ kW × $\frac{30}{60}$ h
= 0.03 kW h

Cost = number of kW h × cost per unit
= 0.03 × 3.5 = sh 0.105

(b) Power = 1 kW. Time = $\frac{30}{60}$ h

Energy consumed = 1 kW × $\frac{30}{60}$ h = 0.5 kW h

Cost = number of kW h × cost per unit
= 0.5 × 3.5 = sh 1.75

Exercise 6.1

1. **(a)** Find out the present cost of one unit of electricity. What other charges are made by the power company for supplying electricity to a domestic consumer?

   (b) Calculate the cost of running each of the appliances listed in Table 6.2 for 1 hour.

2. An electric cooker has an oven rated at 3 kW, a grill at 2 kW and two rings each rated at 500 W. The cooker operates from 240 V mains power.

   (a) Are the different cooking elements likely to be connected in series or in parallel? Give a reason for your answer.

   (b) Would a 20 A fuse be suitable for the cooker, assuming that all parts are switched on? Explain your answer.

   (c) What is the cost of operating all parts for 30 minutes?

3. **(a)** What is the function of a fuse in an electrical circuit?

   (b) Why are the lamps in a domestic circuit usually connected in parallel?

   (c) Earthing is provided for by most three-pin plugs. *(i)* What is the simplest way of earthing power in the home? *(ii)* What purpose is served by earthing?

4. An electric cooker has four elements each rated 4 kW. If the cooker works at full capacity for 30 minutes calculate the cost at 95 cents per unit.

5. **(a)** Power is lost in long-distance power transmission wires. Explain the causes of this power loss and state how each can be minimised.

   (b) Describe three safety devices incorporated in a domestic wiring system and explain how they work.

Summary

- An alternating current or voltage is usually measured by its root mean square (rms) value:

  rms = $\frac{peak\ value}{\sqrt{2}}$ = 0.707 × peak value.

- High-voltage transmission is advantageous because power loss in the lines is small.
- Transmission using a.c. is advantageous because the voltage can easily be stepped up and stepped down by transformers.
- Domestic electricity is supplied as alternating current (a.c.) through the mains at 240 V, 50 Hz.
- The three wires connecting to a standard plug are live (brown or red), neutral (blue or black) and earth (green/yellow or green).
- Switches and fuses must always be placed in the live cable.
- Domestic appliances are connected in parallel to the mains supply.
- A ring main is used to supply sockets – each of the three wires is linked in a complete loop to the power board and therefore carries less current on average.
- 1 kW h is the amount of electrical energy delivered in one hour (3600 s) at a power of 1 kW (1000 W). 1 kW h = 1000 W × 3600 s = 3.6 × 10^6 J.
- 1 kW h is 1 unit.
- Electricity cost = number of kW h × cost per unit.

Review questions

1. **(a)** Describe the main sources of mains electricity in Kenya.
   (b) Discuss the advantages of a.c. power generation as compared to d.c. generation.
   (c) Explain how electric power is transmitted efficiently from the generating station to the consumer.
   (d) What do you understand by 'kilowatt-hour'?

2. **(a)** What is the function a fuse in an electric circuit? What physical property of fuse wire allows it to perform this function?
   (b) Draw a labelled diagram of a correctly wired domestic three-pin plug.

3. **(a)** In wiring a house explain why the lights and power points are wired in parallel while the kilowatt-hour meter is in series. What is the function of the 'earth' wire included in the power points?
   (b) Discuss the possible causes of the following problems:
   (i) no light in one of the rooms,
   (ii) a power failure in the whole house,
   (iii) a power failure affecting a whole town.

4. An electric buzzer has a resistance of 30 Ω and is powered by a 6 V battery.
   (a) How much current is passing through the buzzer circuit?
   (b) How much power is consumed by the buzzer?

5. **(a)** A house has five rooms, each containing one 60 W light bulb. If all the bulbs are left on for two hours how many kilowatt-hours of electrical energy are consumed?
   (b) How much current will a 60 W bulb, operating from a wall outlet of 240 V, draw?
   (c) What is the power used by an air conditioner in a house which draws 7 A of current from a 240 V mains supply?

7 Cathode rays and the cathode ray tube

Learning objectives

After completing the work in this chapter you will be able to:
1. describe the production of cathode rays
2. state the properties of cathode rays
3. explain the functioning of a cathode ray oscilloscope (CRO) and of a television (TV) tube
4. explain the uses of a cathode ray oscilloscope
5. solve problems involving cathode ray oscilloscopes.

7.1 The production of cathode rays

Fig. 7.1 shows the principle of the cathode ray tube. Such tubes were investigated at the end of the nineteenth century when the efficiency of vacuum pumps had advanced to the stage where a very low vacuum could be achieved inside a glass vessel.

The basic cathode ray tube consists of an evacuated glass container with a metal electrode at both ends. In experiments to investigate the conduction of electricity through low-pressure gases in such tubes, it was found that the gas around the negatively charged electrode (the **cathode**) glowed when the pressure was decreased. This glow was described as 'cathode rays'. As the pressure was decreased further the glow vanished, but the opposite end of the tube, beyond the anode, now glowed green. It was as if particles given off from the cathode travel through the gas and strike the end wall and cause it to glow. Fig. 7.2 shows a tube designed to demonstrate that an object placed

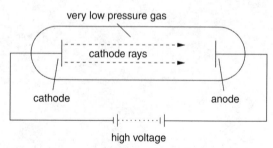

Fig. 7.1 A simple cathode ray tube

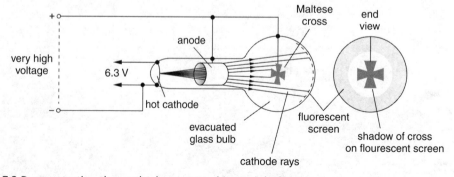

Fig. 7.2 Demonstration that cathode rays travel in straight lines

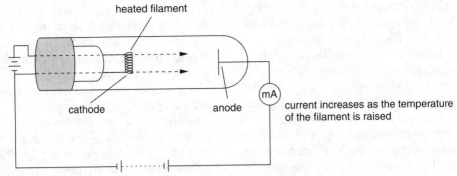

Fig. 7.3 Thermionic emission

in the path of the cathode rays casts a sharp shadow, indicating that cathode rays travel in straight lines.

The properties of cathode rays were investigated extensively. An important discovery was that the production of cathode rays can be greatly increased by heating the cathode to red heat (1000–2000°C). This process is called **thermionic emission** (see Fig. 7.3). The cathode may be made in the form of a wire, like the filament of a light bulb, and heated directly by passing a current through it. Or it may be heated by a separate heating element.

We now understand that cathode rays are made up of **electrons**. The free electrons which carry a current in a metal are held within the metal by their attraction to the positive charges located in the nuclei of the atoms. However, as a result of random collisions an electron may gain sufficient energy to escape from the surface of the metal (just like a molecule evaporating from the surface of a liquid).

If the metal surface is surrounded by gas molecules (as in air) then the electron will not travel far before being absorbed by an ion or deflected back into the metal surface. But if it is in an evacuated tube with a positively charged anode at the other side, the electron can escape from the cathode completely, gaining energy as it is accelerated by its attraction to the positively charged anode.

Heating the cathode wire increases the number of electrons that have sufficient energy to escape from the metal surface, and hence increases the intensity of the cathode rays in the cathode ray tube.

7.2 The properties of cathode rays

Fig. 7.4 shows a cathode ray tube being used to investigate the effect of a magnetic field on cathode rays in order to establish their properties.

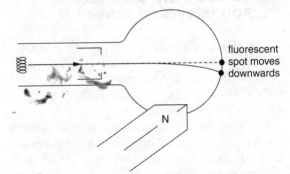

Fig. 7.4 Investigating the properties of cathode rays

Such experiments demonstrate that cathode rays have the following characteristics:
- in the absence of electric or magnetic fields, cathode rays travel from the cathode towards the anode in straight lines;
- cathode rays have kinetic energy;
- cathode rays cause certain substances to fluoresce;
- cathode rays are deflected by a magnetic field in a direction which indicates that they carry a negative charge;
- cathode rays are deflected by an electric field in a direction that indicates that they carry a negative charge;
- cathode rays can produce X-rays on striking matter.

These properties suggest that a beam of cathode rays consists of a stream of tiny negatively charged particles. Through careful measurements of the deflection of the cathode rays by known electric and magnetic fields, the scientist J.J. Thompson was able to calculate the ratio of the charge to the mass of the particles that make up cathode rays. He identified them as electrons – the particles that we now know orbit the nucleus of all atoms and that carry electric current in metals.

In a separate experiment, Robert Millikan measured the charge on the electron by observing the motion of tiny electrically charged oil drops in an electric field. Together, Millikan's and Thompson's results give us values for the charge, e, and the mass of the electron, m_e:

$e = -1.602 \times 10^{-19}$ C
$m_e = 9.11 \times 10^{-31}$ kg

7.3 Cathode ray oscilloscope and television tubes

Those simple cathode ray tubes have been developed for use in devices such as the cathode ray oscilloscope (CRO) and the television (TV) tube. A modern cathode ray tube is illustrated in Fig. 7.5.

The filament (F) and grid (G) together make up the electron gun. The filament is held at a high negative potential relative to earth. Electrons are emitted from the filament by thermionic emission. The strength of the beam (and therefore the brightness of the spot produced on the screen at S) can be adjusted by changing the potential of the grid. If the grid is made more negative than the filament then some electrons are repelled by it and the brightness of the beam is reduced.

Two hollow cylindrical anodes (A) are held at a potential closer to earth potential than the filament and grid. Electrons are thus accelerated towards them by the electric field and the beam gains energy. Although the electrons are attracted to the anodes, their inertia carries them through them to travel on towards the screen. The shape, separation and relative potentials of the anodes are designed to focus the electron beam like a lens, to converge it to a sharp point (S) on the screen. The screen is coated with a fluorescent material that glows when the electron beam falls on it.

In a CRO tube, the beam passes between two sets of parallel metal plates on its way to the screen – these are called the deflecting plates. An electric potential applied between the Y plates deflects the beam vertically; a potential between the X plates deflects the beam horizontally.

In a TV tube deflection of the beam is produced by magnetic fields rather than electric fields. Varying currents are applied to four electromagnets placed around the tube to deflect the beam horizontally and vertically to any point on the screen.

The cathode ray oscilloscope

A CRO, such as that shown in Fig. 7.6, is a measuring instrument based on the cathode ray tube. It is designed to display and measure electrical signals applied to its input terminals. Amplifiers in the oscilloscope adjust the voltages of input signals to appropriate levels and then apply them to the deflection plates. Signals may be applied independently to the X

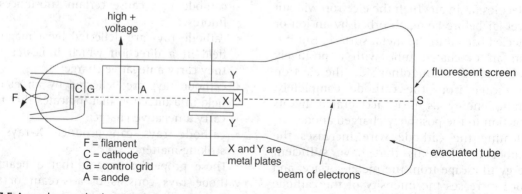

Fig. 7.5 A modern cathode ray tube

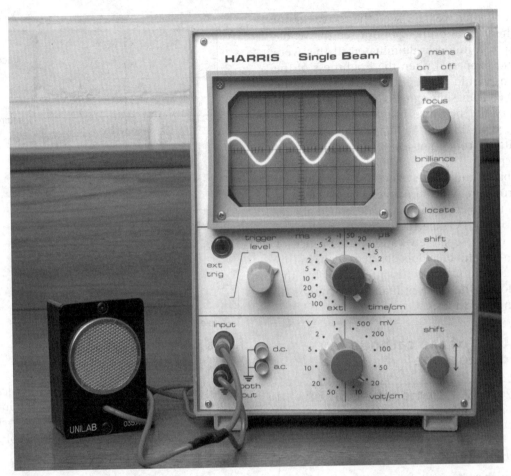

Fig. 7.6 A cathode ray oscilloscope

and Y inputs to deflect the beam horizontally or vertically. The electron beam acts as a very low inertia pointer or needle which can follow very rapid changes in the voltages being measured. Details of how the oscilloscope is used to display and measure different types of signal, are given in Section 7.4.

The television tube

In a TV a cathode ray tube is used to display images. To achieve this, scanning signals are applied to the deflection coils so that the beam is constantly scanned in a series of horizontal lines to and fro across the entire screen. The path followed by the beam is indicated in Fig. 7.7 – this path is described as a raster scan. The beam follows this path once every one-twentyfifth of a second. Each point on the screen is illuminated by the beam successively but (because the fluorescence of

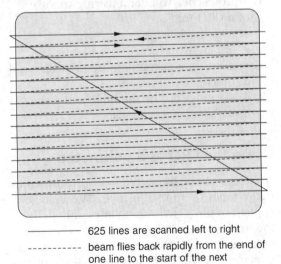

——— 625 lines are scanned left to right
- - - - - - beam flies back rapidly from the end of one line to the start of the next

Fig. 7.7 TV scan

the coating material persists for a short time after the beam has passed, and images on the retina of the eye persist for up to one-tenth of a second after we have seen them) we do not see the motion of the spot but see the whole screen continuously illuminated. If this were all that happened, then the brightness of the screen would be uniform. However, the received television signal is applied to the grid to vary the brightness of the beam as it is scanned. In this way an image which varies in brightness from point-to-point is projected on the screen.

A black and white television has just one electron gun, which produces a single electron beam to project the image as described above. In a colour TV tube there are three electron guns which separately cause fluorescence of red, green and blue dots on the screen. In this way a colour image is produced by the addition of the three primary light colours in varying intensities.

Uses of the cathode ray oscilloscope

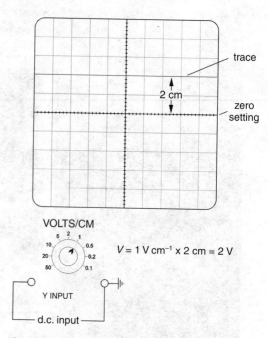

Fig. 7.8 A cathode ray oscilloscope used as a voltmeter

$V = 1\ \text{V cm}^{-1} \times 2\ \text{cm} = 2\ \text{V}$

Activity 7.1 Demonstration of the cathode ray oscilloscope in use

Method

1. Your teacher will demonstrate the use of a CRO to:
   - measure voltage,
   - display waveforms,
   - measure and compare frequencies,
   as described in Section 7.4.
2. Make notes on the methods and write down your observations.

Using the CRO as a voltmeter

A voltage can be measured by connecting it to the Y input of an oscilloscope. The voltage is applied to the Y deflection plates, and deflects the oscilloscope beam vertically. The sensitivity of the CRO is set with the *volt/cm* selector. On the CRO shown in Fig. 7.8 the sensitivities range from 0.1 V cm^{-1} to 50 V cm^{-1} and it shows the effect of connecting a 2 V d.c signal to the Y input when the voltage selector is at 1 V cm^{-1}. The time base signal (see below) causes the d.c. signal to appear as a horizontal line on the screen. The line moves vertically up or down depending on whether the input signal is positive or negative. The Y-shift control (a knob coaxial with the voltage

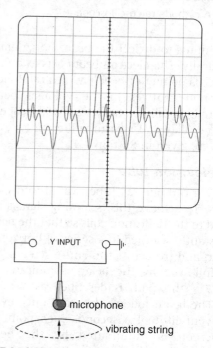

Fig. 7.9 Displaying a waveform

sensitivity setting) can be used to set the position of the trace for zero voltage input.

Using the CRO to display a waveform

The CRO can also be used to display voltage as a function of time and the screen displays a waveform. Fig. 7.9 shows the display on a CRO when a microphone is connected to the Y-input and a musical sound (a plucked string) reaches the microphone. The microphone produces a voltage which varies in the same way as the pressure variations of the sound wave. The CRO displays this waveform, which is, in effect, a graph of voltage against time.

Time base

In order to display a waveform, a signal called a **time base** is connected to the X-deflection plates of the cathode ray tube. This signal is generated internally within the oscilloscope. Fig. 7.10 shows the time base signal.

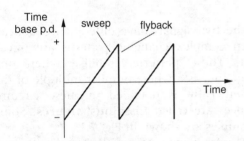

Fig. 7.10 The time base signal

Its shape is described as a sawtooth, for obvious reasons. The sweep region of the time base signal causes the spot to move at a constant speed from left to right across the oscilloscope screen – the flyback region returns the spot to the start of the trace and the sweep is repeated. With this signal causing the spot to sweep horizontally, the external signal (from the microphone for example), connected to the Y-deflection plates, causes the spot to move vertically up and down as it sweeps horizontally to and fro. In this way the waveform of a signal is traced out.

The time base control on the CRO allows the user to adjust the speed at which the spot sweeps. On the oscilloscope shown in Fig. 7.6 the time base settings vary from $100\,\text{ms}\,\text{cm}^{-1}$ (the slowest sweep speed or longest time base) to $1\,\mu\text{s}\,\text{cm}^{-1}$ (the fastest sweep speed or shortest time base). The user adjusts the time base according to the frequency of the signal being investigated to obtain a suitable display of the waveform. If the time base is too long then many cycles of the wave will be displayed on the screen simultaneously and the shape of an individual cycle will not be clear. If the time base is set too short then only a fraction of a period will be displayed and the complete waveform will not be visible.

Fine adjustments to the time base can be made with the variable adjustment knob which is coaxial with the main time base selector. For the time base readings on the main selector to be correct this knob must be rotated fully clockwise to the calibrated position.

Triggering

For a stable trace to be produced on the screen, the waveform must be plotted in exactly the same location for each sweep of the time base signal. This is achieved by adjusting the trigger control. When triggering is used, each sweep starts only when the input signal rises to a certain value (the trigger level) which is adjusted with the trigger control. Consecutive sweeps are thus all triggered at the same point on the waveform and superimpose one on top of another.

Using the CRO to compare the phase and frequency of two signals

Two sine wave signals can be compared by connecting one of them to the X-input and the other to the Y-input of the oscilloscope. When this is done, the time base signal is turned off by setting the time base control to 'ext' (external X-input).

Fig. 7.11 shows the CRO trace for two signals with the same frequency but different relative phases. If the two signals have the same frequency, amplitude and phase then the trace is a diagonal line – think about the path that each signal independently makes the trace follow, the resulting trace is the vector sum of the two independent traces as shown in Fig. 7.11(a). If the two signals have the same frequency and amplitude, but a phase difference of $\frac{\pi}{2}$, then the trace is a perfect circle as shown in Fig. 7.11(b).

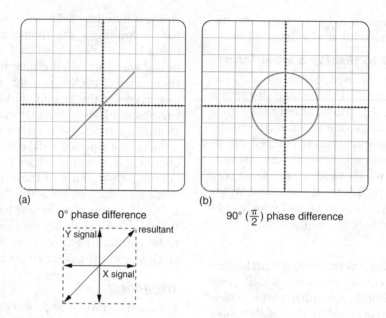

Fig. 7.11 Comparing phase
(a) No difference in phase; (b) Phase difference of 90°

Example

Can you predict the shape of the trace for the following combinations of signals?
(a) Equal amplitude and frequency; phase difference = π.
(b) Equal amplitude and frequency; phase difference = $\frac{\pi}{4}$.
(c) Equal amplitude and frequency; phase difference = $\frac{3\pi}{4}$.

Answer

The shapes of the traces are shown in Fig. 7.12.

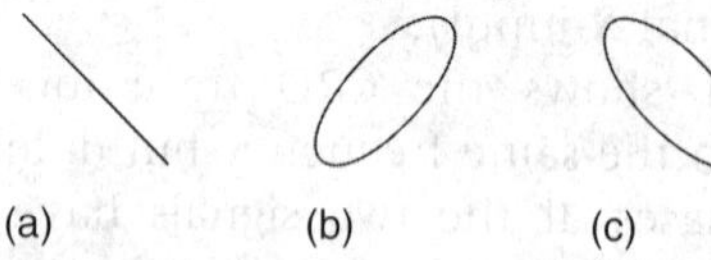

If the two signals have different frequencies then complex changing patterns are traced out. These patterns stabilise when one frequency is a whole number multiple of the other. Patterns obtained in these circumstances are called **Lissajous' figures**. Some examples are shown in Fig. 7.13.

The frequency ratio of the two frequencies can be determined by counting the number of loops in the Lissajous' figures. A general rule for determining this frequency ratio is illustrated in Fig. 7.14. The ratio is given by

$$\frac{f_Y}{f_X} = \frac{\text{number of loops touching horizontal line}}{\text{number of loops touching vertical line}}$$

Fig. 7.12

Fig. 7.13 Lissajous' figures

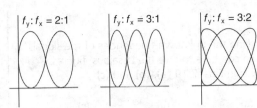

Fig. 7.14 Using Lissajous' figures to determine a frequency ratio

7.5 Solving oscilloscope problems

Example 1

What are **(a)** the frequency and **(b)** the amplitude of the signal shown in Fig. 7.15?

Answer

(a) Time base = 50 μs cm^{-1}
One period, T, of the signal occupies 4.0 cm, therefore $T = 4.0 \times 50 = 200$ μs
$= 2.0 \times 10^{-4}$ s

$f = \dfrac{1}{T} = \dfrac{1}{2.0 \times 10^{-4}} = 5000$ Hz $= 5.0$ kHz.

(b) Sensitivity
= 0.2 V cm^{-1}; amplitude on screen
= 2.5 cm, therefore amplitude
= 2.5 cm × 0.2 V cm^{-1} = 0.5 V

Example 2

A signal with frequency 500 Hz is connected to the X-input of an oscilloscope. An unknown signal is connected to the Y-input. Fig. 7.16

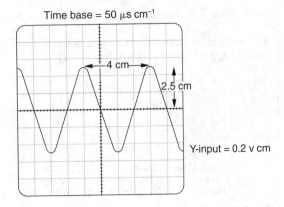

Fig. 7.15

shows the resulting trace. What is the frequency of the signal?

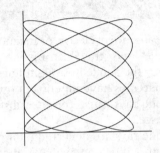

Fig. 7.16

Answer
Number of loops touching horizontal line = 3; number of loops touching vertical line = 5.

$\dfrac{f_Y}{f_X} = \dfrac{3}{5}$ therefore $f_Y = \dfrac{3}{5} \times f_X = \dfrac{3}{5} \times 500 = 300$ Hz.

Exercise 7.1

1. Fig. 7.17**(a)** shows an oscilloscope trace when there is no Y-input. What are the d.c. voltages of the input signals connected in **(b)**, **(c)** and **(d)**?

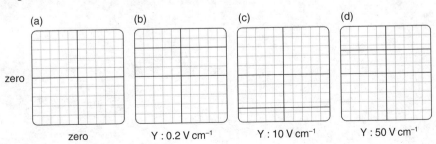

Fig. 7.17

(continued)

Exercise 7.1 (continued)

2. Find the frequencies and amplitudes of the signals shown in Figs. 7.18 **(a)**, **(b)** and **(c)**.

3. Find the unknown frequencies of the signals which combine to give the Lissajous' figures shown in Figs. 7.19 **(a)**, **(b)** and **(c)**.

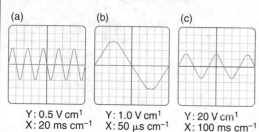

(a) Y: 0.5 V cm⁻¹ X: 20 ms cm⁻¹
(b) Y: 1.0 V cm⁻¹ X: 50 μs cm⁻¹
(c) Y: 20 V cm⁻¹ X: 100 ms cm⁻¹

Fig. 7.18

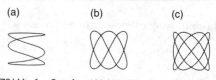

(a) $f_X = 70\,\text{kHz}, f_Y = ?$
(b) $f_X = 100\,\text{kHz}, f_Y = ?$
(c) $f_X = ?, f_Y = 4\,\text{kHz}$

Fig. 7.19

Summary

- In the absence of electric or magnetic fields, cathode rays travel from the cathode towards the anode in straight lines.
- Cathode rays have kinetic energy.
- Cathode rays cause certain substances to fluoresce.
- Cathode rays are deflected by a magnetic field in a direction which indicates that they carry a negative charge.
- Cathode rays are deflected by an electric field in a direction that indicates that they carry a negative charge.
- Cathode rays can produce X-rays when they strike matter.
- Cathode rays are made up of electrons – an electron has charge $e = -1.602 \times 10^{-19}$ C and mass $m_e = 9.11 \times 10^{-31}$ kg.

- The cathode ray oscilloscope is a measuring instrument based on the cathode ray tube.
- A sawtooth time base signal causes the spot to move at a constant speed from left to right across the oscilloscope screen, and then fly back to the start.
- Lissajous' figures are produced when two sinusoidal signals, with a whole number frequency ratio, are applied to the X- and Y-inputs of an oscilloscope.
- $\dfrac{f_Y}{f_X} = \dfrac{\text{number of loops touching horizontal line}}{\text{number of loops touching vertical line}}$
- In a television, a cathode ray tube is used to display images. Scanning signals are applied to deflection coils to scan the beam in a series of horizontal lines to and fro across the screen.

Review questions

1. **(a)** What are cathode rays?
   (b) List the properties of cathode rays.
   (c) Draw a labelled diagram showing the components of a simple cathode ray tube used to investigate their properties.

2. **(a)** Draw a sketch diagram to show the main features of the cathode ray tube of a cathode ray oscilloscope.
   (b) With the help of your sketch diagram, describe how the electrons are produced in the tube.
   (c) State and explain the function of the grid.
   (d) Describe how the deflection system of a TV differs from that of a CRO.

3. With the aid of a suitable sketch of the waveform on the oscilloscope screen, describe how a cathode ray oscilloscope is used to determine the amplitude and frequency of a sinusoidal signal applied to the Y-input.

8 X-rays

Learning objectives

After completing the work in this chapter you will be able to:

1. explain the production of X-rays
2. state the properties of X-rays
3. state the dangers of X-rays
4. explain the uses of X-rays
5. solve numerical problems involving X-rays.

8.1 The production of X-rays

X-rays were discovered by William Röntgen in 1895. He called them 'X-rays' because at first their nature was unknown. Röntgen discovered that invisible radiation was emitted from the end of a cathode ray tube which the cathode rays (electrons) were striking. This radiation was detected by its effect on a photographic film. Although the radiation could not be seen, it darkened the film in a similar way to light, but unlike light it was able to pass through the opaque wrapping to affect the film beneath.

X-rays are emitted whenever cathode rays are brought to rest by matter. The kinetic energy that the cathode rays gain in accelerating between the cathode and anode of a cathode ray tube is transformed into X-rays when the cathode rays are brought back to rest. We now know that X-rays are high frequency electromagnetic radiation (see Chapter 4). The frequencies of X-rays are in the range $10^{16} - 10^{21}$ Hz and their wavelengths in the range $10^{-7} - 10^{-13}$ m.

Fig. 8.1 shows a cross-section of an X-ray tube. Cathode rays emitted by a hot filament are accelerated towards a copper anode held at a potential difference of several tens of thousand volts with respect to the cathode.

Another heavy metal, such as lead, is sometimes inserted into the anode at the point where the X-rays strike in order to produce X-rays with characteristic energies. When cathode rays strike the heavy-metal target most of their energy is turned into heat, but about 0.5% is ultimately converted into electromagnetic radiation in the X-ray waveband. The excess heat energy produced is carried away by the heavy copper anode, which may have water cooling coils wrapped around it.

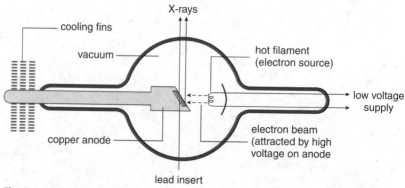

Fig. 8.1 An X-ray tube

8.2 Energy changes in an X-ray tube

The cathode rays that excite X-ray emission in the X-ray tube gain kinetic energy as they are accelerated from the cathode to the anode by the large potential difference applied. The potential difference between two points is defined as the energy change per unit charge in moving between those two points. If the potential difference between the cathode and the anode is V, then the energy gained by an electron with charge e in accelerating from one to the other is eV. This is the kinetic energy of the electron as it strikes the anode,

$$eV = \tfrac{1}{2} mv^2.$$

In the next chapter it will be shown that the energy of a photon of electromagnetic radiation is given by $E = hf$, where h is a constant called the Planck constant and f is the frequency of the radiation. A photon is the smallest quantity (a quantum) of electromagnetic radiation of a given frequency.

Thus, if one electron gives all its kinetic energy to one X-ray photon then, from the conservation of energy,

$$hf = \tfrac{1}{2} mv^2 = eV$$

This is the maximum possible energy for the X-ray photons produced by cathode rays accelerated through a voltage V. The maximum frequency of the X-rays produced by an X-ray tube operated at V volts is thus

$$f_{max} = \frac{eV}{h}.$$

Using the wave equation, $c = f\lambda$, the minimum wavelength of the X-rays emitted is thus

$$\lambda_{min} = \frac{c}{f_{max}} = \frac{ch}{eV}.$$

X-rays with this frequency and wavelength are the most energetic emitted by the X-ray tube. Most of the radiation emitted by the tube will have a lower frequency and a longer wavelength than these limiting values, f_{max} and λ_{min}, since the photons will not carry all the energy originally given to an electron by the accelerating voltage. X-rays with higher frequencies and shorter wavelengths, which carry more energy per photon, may be generated by increasing the voltage applied to the X-ray tube.

Intensity

The brightness, or intensity, of the X-ray beam is controlled by the *quantity* of the X-ray photons produced, not their energy. This quantity depends on the current (the number of electrons passing from the cathode to the anode per second) in the tube. But increasing the current incurs costs – the anode must be cooled more effectively or it will overheat and melt. High intensity X-rays tubes have cylindrical rotating anodes so that cathode rays do not constantly strike the same point on the metal surface. Also, a large moving anode has a higher heat capacity than a small stationary anode.

Activity 8.1 Understanding energy changes in an X-ray tube

You will need:
- materials for preparing a poster.

Method
1. Working in a group, copy the diagram of an X-ray tube in Fig. 8.1 onto your poster.
2. Review Section 8.2 which outlines the energy changes that take place in the X-ray tube.
3. On your poster, indicate the motion of the electrons from the cathode to the anode in the X-ray tube and the resulting emission of the X-ray photons from the anode.
4. Label your poster to show the energies of the electrons/photons at the various stages of the X-ray production process.
5. Write a brief caption for the poster summarising the energy changes that take place.

8.3 The properties of X-rays

The basic properties and uses of X-rays were outlined in Chapter 4. In summary:
- they travel, like all types of electromagnetic radiation, in straight lines at the speed of light, c, which is 3×10^8 m s^{-1} in a vacuum;
- they exhibit diffraction and interference when passed through regular crystalline structures – they therefore behave as waves (see Fig. 8.2);

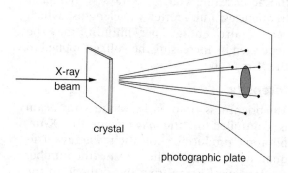

Fig. 8.2 X-ray diffraction: when X-rays pass through a crystal they are diffracted by the regularly spaced atoms

- they are not deflected by electric or magnetic fields, and therefore carry no charge;
- they cause some materials to fluoresce and affect photographic plates;
- they cause ionisation in gases;
- they cause metals to emit electrons (photoelectric emission);
- they penetrate matter significantly – for example, they can pass through human flesh, paper and other materials;
- they are absorbed by metals and may be completely absorbed by a thin layer of lead.

Low-energy X-rays are described as *soft* X-rays – they penetrate low density materials such as flesh. Higher energy X-rays are described as *hard* X-rays and can penetrate metal or stone.

8.4 The dangers of X-rays

Like other ionising radiations, excessive or prolonged exposure to X-rays can damage living organisms. The types of damage induced include:
- radiation burns
- radiation sickness
- radiation induced cancers
- genetic damage.

X-rays may be used safely at low doses for applications such as airport safety checks, and industrial and medical X-ray photography.

Precautions

However, appropriate safety precautions must always be observed:
- the X-ray dose must be kept to a minimum;
- X-ray sources must be appropriately screened with lead shielding;
- operators of X-ray equipment must be appropriately trained and wear personal radiation monitors.

Personal radiation monitors often take the form of a 'film badge' – a piece of radiation-sensitive photographic film which records the quantity of radiation to which the wearer is exposed during the course of their work (see Fig. 8.3). Film badges are checked at regular intervals, and strict limits are set on the exposure levels allowed to ensure the safety of the radiation workers.

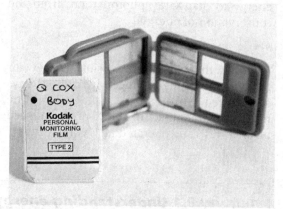

Fig. 8.3 A film badge used to monitor personal radiation exposure

8.5 Uses of X-rays

X-rays can be used:
- to diagnose fractured or displaced bones in the body (see Fig. 8.4);
- to locate unusual objects in the body, for example a swallowed safety-pin;
- to find flaws in metal castings and welded joints, for example in pipelines;

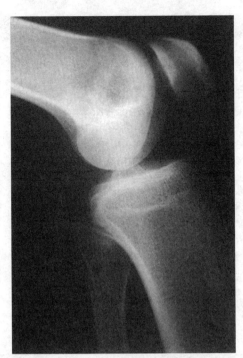

Fig. 8.4 An X–ray photograph of a human knee showing the thigh bone (femur) and the two bones of the lower leg (tibia and fibula).

- to examine machines without taking them apart;
- to destroy cancer cells (although exposure to X-rays can also cause cancer);
- to study crystal structures;
- to perform airport security checks.

8.6 Solving problems on X-rays

Example

An X-ray tube is operated at 40 kV. Calculate:
(a) the kinetic energy of an electron accelerated by this potential,
(b) the speed to which an electron is accelerated by this potential,
(c) the maximum frequency of the X-rays produced by the tube,
(d) the minimum wavelength of the radiation. (velocity of light $c = 3 \times 10^8$ m s^{-1}; charge on the electron $e = 1.602 \times 10^{-19}$ C; mass of the electron $m_e = 9.11 \times 10^{-31}$ kg; Planck constant $h = 6.63 \times 10^{-34}$ J s)

Answer

(a) KE $= eV = 1.602 \times 10^{-19} \times 40 \times 10^3$
$= 6.41 \times 10^{-15}$ J.

(b) KE $= \frac{1}{2}mv^2$

So $v = \sqrt{\frac{2 \times \text{KE}}{m}} = \sqrt{\frac{2 \times 6.41 \times 10^{-15}}{9.11 \times 10^{-31}}}$
$= 1.19 \times 10^8$ m s^{-1}

(c) $f_{max} = \frac{eV}{h} = \frac{1.602 \times 10^{-19} \times 40 \times 10^3}{6.63 \times 10^{-34}}$
$= 9.67 \times 10^{18}$ Hz.

(d) $\lambda_{min} = \frac{c}{f_{max}} = \frac{3 \times 10^8}{9.67 \times 10^{18}} = 3.10 \times 10^{-11}$ m.

Exercise 8.1

1. Draw a labelled diagram of an X-ray tube. What features of the operation of the tube determine:
   (a) the intensity of the X-rays,
   (b) the penetrating power (maximum energy) of the X-rays?

2. Outline the energy changes that take place in an X-ray tube.

3. An accelerating potential of 25 kV is applied to an X-ray tube. Calculate:
   (a) the kinetic energy of the electrons accelerated by this potential,
   (b) the speed to which an electron is accelerated by this potential,
   (c) the maximum frequency of the X-rays produced by the tube,
   (d) the minimum wavelength of the radiation.
   (e) Explain how X-rays with a shorter wavelength could be produced.

Summary

- X-rays are produced in an X-ray tube when cathode rays strike a heavy-metal target.
- The maximum frequency of the X-rays produced is given by $f_{max} = \dfrac{eV}{h}$.
- The minimum wavelength is given by $\lambda_{min} = \dfrac{c}{f_{max}}$.
- X-rays travel in straight lines, like other forms of electromagnetic radiation, at the speed of light, $c = 3 \times 10^8$ m s^{-1} in a vacuum.
- X-rays exhibit diffraction and interference when passed through regular crystalline structures – they therefore behave as waves.
- X-rays are not deflected by electric or magnetic fields, and therefore carry no charge.
- X-rays cause some materials to fluoresce and affect photographic plates.
- X-rays cause ionisation in gases.
- X-rays cause metals to emit electrons (photoelectric emission).
- X-rays penetrate matter significantly – for example, X-rays can pass through human flesh, paper and other materials.
- X-rays are absorbed by metals and can be completely absorbed by a thin layer of lead.

Review questions

1. **(a)** Draw an X-ray tube, label its major components and briefly describe its operation.
   (b) Give a qualitative description of the energy changes that take place in the tube when X-rays are produced.

2. **(a)** List the properties of X-rays.
   (b) Outline three uses of X-rays.
   (c) Discuss the hazards created by X-rays and describe the precautions that should be taken in their use.

3. **(a)** A W-shaped metal plate is placed on top of a black light-tight envelope containing a sheet of photographic film. An X-ray unit is placed about 2 cm above the film and switched on. What would be your observation when the film is developed? Explain your answer.
   (b) X-rays are incident on a negatively charged electroscope. Describe and explain what happens.
   (c) X-rays are directed through the air in the space between two parallel metal plates. The plates are connected in series with a voltage source and a galvanometer. A current is registered by the galvanometer. Explain this observation.

4. An X-ray tube accelerates electrons through a p.d. of 25 kV. Measurements show that X-ray radiation produced from the target has a minimum wavelength of 49.5 pm. Calculate
   (a) the maximum energy of the X-ray photons,
   (b) the maximum frequency of the X-rays,
   (c) a value for Planck's constant.
   ($c = 3 \times 10^8$ m s^{-1}, $e = 1.6 \times 10^{-19}$ C)

9 The photoelectric effect

Learning objectives

After completing the work in this chapter you will be able to:

1. perform and describe simple experiments to illustrate the photoelectric effect
2. explain the factors affecting photoelectric emission
3. apply the equation $E = hf$ to calculate the energy of photons
4. define threshold frequency, work function and the electron volt
5. explain photoelectric emission using the Einstein equation, $hf = hf_0 + \frac{1}{2}mv^2$.
6. explain the applications of the photoelectric effect
7. solve numerical problems involving photoelectric emissions.

9.1 The photoelectric effect

Electromagnetic radiation carries energy. When electromagnetic rays are incident on a metal surface, some electrons in the metal may gain enough energy to escape from the surface. This is why a negatively charged conductor can be discharged if light of sufficient energy falls on it. In a photocell the emitted electrons are collected by an anode to produce an electric current, converting light energy into electricity. The emission of electrons by a metal exposed to light is called the **photoelectric effect**. Study of the photoelectric effect has led to important developments in our understanding of the way in which energy is carried by electromagnetic waves.

Activity 9.1 Experiments to demonstrate and investigate the photoelectric effect

You will need:
- an electroscope
- a zinc plate
- some emery paper
- a glass rod and a silk cloth
- an ebonite rod and a piece of fur
- a table lamp
- an ultraviolet lamp
- a sheet of glass.

Method

1. Clean the surface of the zinc plate with emery paper and lay it on the electroscope cap.
2. Rub the glass rod with silk and charge the electroscope positively.
3. Shine light from the table lamp on the zinc plate, and observe if the electroscope leaf falls or rises.
4. Repeat steps 2 and 3 using the ultraviolet lamp, and observe if the electroscope leaf falls or rises.
5. Discharge the electroscope and lay the zinc plate on it again.
6. Rub the ebonite rod with fur and charge the electroscope negatively.
7. Shine light from the table lamp on the zinc plate, making the light as intense as possible by holding it as close as possible to the surface, and observe if the electroscope leaf falls or rises.

(continued)

Activity 9.1 (continued)

8. Repeat steps 6 and 7 using the ultraviolet lamp, and observe if the electroscope leaf falls or rises.
9. Recharge the electroscope negatively and try to discharge it by shining ultraviolet light on the zinc plate (see Fig. 9.1). As the leaf begins to fall, insert a sheet of glass between the lamp and the plate. What happens?

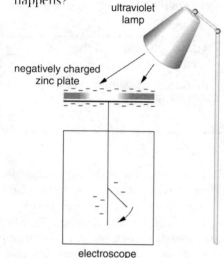

Fig. 9.1 Demonstrating the photoelectric effect

Observations
In this experiment you should observe that:
- neither visible light nor ultraviolet light discharges a positively charged zinc plate;
- a negatively charged plate is not discharged by visible light, but is discharged by ultraviolet light;
- inserting glass between the ultraviolet lamp and the plate halts the discharge (the electroscope leaf ceases to fall), and removing the glass plate allows the discharge to proceed.

Discussion
The observation that a negatively charged plate is discharged, while a positively charged plate is not, suggests that the effect of the ultraviolet light is to eject electrons (which carry a negative charge) from the metal surface. Ultraviolet light carries more energy than visible light and so the fact that visible light does not discharge the plate suggests that a certain minimum energy is required to eject the electrons.

A sheet of glass stops the discharging process because glass absorbs ultraviolet light.

Understanding the photoelectric effect

The photoelectric effect was not fully understood until Albert Einstein used the **quantum theory** of electromagnetic radiation to explain the observations, and the success of his explanation helped to demonstrate that his quantum theory was correct. In the quantum theory, light and other forms of electromagnetic waves are understood to travel as photons.

Photons

Water waves and sound waves are continuous waves, like those shown in Fig. 9.2(a). A continuous wave can carry any amount of energy depending on its amplitude and frequency. There is no minimum value for the energy of the wave – the energy may be reduced simply by decreasing the amplitude of the wave.

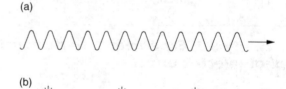

Fig. 9.2
(a) A continuous wave
(b) Photons (wave packets)

Quantum theory, however, shows that electromagnetic waves travel as wave packets or **photons**, as illustrated in Fig. 9.2(b). Each photon carries a definite amount of energy, which is proportional to the frequency, f, of the radiation. The energy of a photon is given by $E = hf$.

The Planck constant

The constant h in the equation for the energy of a photon is called the **Planck constant**. It has the value 6.63×10^{-34} J s. As you will see in Section 9.4, this value can be determined by measurement of the photoelectric effect.

Example

(a) What is the energy of a photon of visible light with frequency 5.0×10^{14} Hz?
(b) What is the energy of an X-ray photon with frequency 10^{18} Hz?

Answer
(a) $E = hf = 6.63 \times 10^{-34} \times 5.0 \times 10^{14}$
$= 3.3 \times 10^{-19}$ J.
(b) $E = hf = 6.63 \times 10^{-34} \times 10^{18}$
$= 6.63 \times 10^{-16}$ J.

Threshold frequency

Experiments such as those in Activity 9.1 show that radiation below a certain frequency does not eject electrons from a metal surface, no matter how intense the light. This frequency is called the **threshold frequency**, f_0. As you observed in the activity, visible light does not eject electrons from zinc, but ultraviolet light (with higher frequency) does. This is illustrated in Fig. 9.3.

If visible light is a continuous wave, the existence of the threshold frequency cannot be understood. If the visible light is made bright enough it should give the electrons enough energy to escape – but this does not happen.

Einstein was able to use the photon model of light to explain the threshold frequency. He reasoned that a certain minimum amount of energy was needed to eject an electron from a metal surface. If a single electron can gain energy only by absorbing the energy of a single photon, then the electron will only escape if the energy of the photon exceeds the minimum required. None of the photons of visible light have sufficient energy to eject an electron, and therefore photoemission does not occur for light of this frequency. However, ultraviolet photons have a higher frequency and therefore each photon carries a higher energy – so photons of ultraviolet light can eject electrons. As the frequency of the incident light is increased there is a threshold frequency at which photoelectric emission takes place.

Work function

The work function, W, is the minimum amount of energy needed to eject an electron from a metal surface. It is the work needed to remove a negatively charged electron from the metal, overcoming the force of attraction to the positively charged nuclei of the metal atoms. The value of the work function differs from one metal to another.

Electron volt

When dealing with energy changes on the atomic scale, for example the energy changes of electrons in the photoelectric effect, the joule is a very large unit. The work function of zinc, for example, is just 6.93×10^{-19} J. It is convenient to use a much smaller energy unit called the **electron volt** to express these small energy values. You will recall that the definition of the volt is the potential difference between two points such that the energy

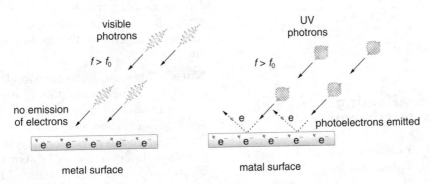

Fig. 9.3 The threshold frequency

transferred when one coulomb of charge is moved between those points is one joule (1 V = 1 J C^{-1}). The energy change, E, when q coulombs are moved through V volts is thus $E = qV$.

The energy change when an electron which carries charge e is moved through V volts is thus eV joules. The electron volt is defined as the energy change when one electron is moved through one volt:
1 eV = 1.602 × 10^{-19} C × 1 V = 1.602 × 10^{-19} J

Example

(a) Express the work function of zinc in electron volts.
(b) The work function of potassium is 2.2 eV. Express this energy in joules.
(c) In a cathode ray tube, an electron is accelerated from the cathode to the anode by a potential difference of 2000 V.
  (i) Calculate the energy gained by the electron in electron volts and joules.
  (ii) Calculate the speed of the electron as it reaches the anode.

Answer
(a) 1 eV = 1.602 × 10^{-19} J
Therefore 6.93 × 10^{-19} J
$= \frac{6.93 \times 10^{-19}}{1.602 \times 10^{-19}}$ eV = 4.33 eV.

(b) 1 eV = 1.602 × 10^{-19} J
Therefore 2.2 eV = 2.2 × 1.602 × 10^{-19} J
= 3.5 × 10^{-19} J.

(c) (i) Energy gained by electron in accelerating through 2000 V = 2000 eV.
1 eV = 1.602 × 10^{-19} J
Therefore 2000 eV = 2000 × 1.602 × 10^{-19} J = 3.204 × 10^{-16} J.

(ii) KE = $\frac{1}{2}mv^2$
So $v = \sqrt{\frac{2 \times KE}{m}} = \sqrt{\frac{2 \times 3.204 \times 10^{-16}}{9.11 \times 10^{-31}}}$
= 2.65 × 10^7 m s^{-1}.

9.2 Factors affecting photoelectric emission

Fig. 9.4 shows apparatus that can be used to investigate the factors affecting photoelectric emission in detail.

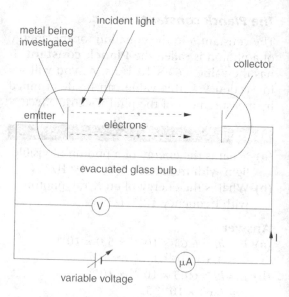

Fig. 9.4 Apparatus used to investigate the photoelectric effect

The metal to be investigated is made one of the electrodes in an evacuated glass bulb. Light of different frequencies is shone on the metal surface. The electrons emitted from the surface travel through the vacuum to the second electrode, creating a current which is indicated on the microammeter. The current reading is thus a measure of the number of electrons being emitted.

The energy of the electrons can be found by adjusting the potential between the two metal plates. If the second plate is made negative with respect to the first, electrons are repelled from it. Only those that start with sufficient energy will reach the second electrode – the current will fall as fewer electrons reach it. As the negative voltage is increased, the current eventually drops to zero, as shown in Fig. 9.5. The voltage at which the current becomes zero (the stopping potential, V_s) gives the maximum energy of the emitted electrons in electron volts.

Note: This can be understood from the principle of the conservation of energy. If the electrons have maximum energy V_s electron volts as they leave the metal surface, then they will lose this energy in moving through a potential difference of $-V_s$ volts to reach the second electrode. If the voltage of the second electrode is decreased any further (made more

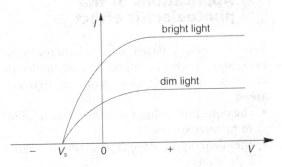

Fig. 9.5 Current against applied voltage for the photoelectric effect

negative), none of the electrons will have sufficient energy to reach it.

Experiments with this apparatus show that the factors that affect the photoelectric effect are as follows.
- Electrons are only emitted when the frequency of the incident light exceeds a certain threshold frequency.
- Different metals have different threshold frequencies (see Fig 9.6).
- Below the threshold frequency there is no photoemission, no matter how bright the light.
- Above the threshold frequency photoemission takes place no matter how dim the light.
- Above the threshold frequency the energy of the emitted electrons increases with increasing frequency of the light (see Fig. 9.6).
- The energy of the electrons does not depend on the brightness (intensity) of the light – increasing the intensity increases the number of electrons emitted, not their energy.

9.3 Energy of photons

As we have seen, the dependence of the photoelectric effect on light frequency and intensity can be explained if light (and other forms of electromagnetic radiation) is made up of photons that carry energy in fixed amounts determined by the frequency of the radiation. The energy of a single photon is given by the expression $E = hf$.

In the photoelectric effect, if the threshold frequency is exceeded, then even a single photon has sufficient energy to eject an electron from the metal surface. This explains why the photoelectric effect occurs even in very dim light – so long as its frequency is greater than the threshold frequency. However, if the frequency is below the threshold frequency then, no matter how bright the light, none of the photons has

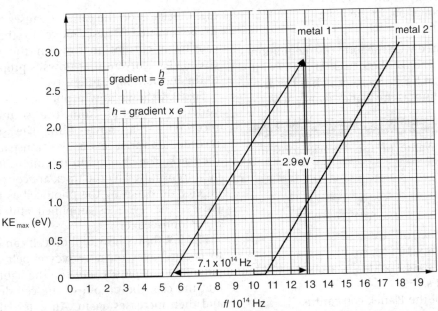

Fig. 9.6 Graphs of maximum electron energy (V_s) against incident light frequency (f) for two different metals

sufficient energy to eject an electron from the metal surface, and the photoelectric effect does not take place.

9.4 The Einstein equation

Einstein's equation gives the energy of the electrons emitted during the photoelectric effect. It is derived as follows. Suppose the work function of the metal surface is W. The threshold frequency, f_0, for light that will just eject an electron from the surface is given by
$hf_0 = W$.
The energy of electrons ejected by light of frequency f, which must be greater than the threshold frequency, will thus be
$E = hf - W = hf - hf_0$.
This energy is the kinetic energy of the emitted electrons, so
$\frac{1}{2}mv^2 = hf - hf_0$.
Therefore $hf = hf_0 + \frac{1}{2}mv^2$

This is **Einstein's equation** for the photoelectric effect. Einstein's equation may be understood in terms of the conservation of energy. The energy of the incident photon, hf, is transferred to the work done in removing the electron from the metal surface, hf_0, plus the kinetic energy of the electron.

Einstein's equation, in the form
$\frac{1}{2}mv^2 = hf - hf_0$
can be compared with the form of an equation for a straight line. Looking back at Fig. 9.6, it can be seen that $KE_{max} = hf - hf_0$. Thus the gradient of both lines is the Planck constant, h.

Example

Estimate the value of the Planck constant from Fig. 9.6.

Answer
Gradient = $\dfrac{\text{rise}}{\text{run}} = \dfrac{2.9 \text{ eV}}{7.1 \times 10^{14} \text{ Hz}}$

  $= 4.1 \times 10^{-15}$ eV s.
Now 1 eV = 1.6×10^{-19} J, so
4.1×10^{-15} eV s = $4.1 \times 10^{-15} \times 1.6 \times 10^{-19}$ J s
$= 6.6 \times 10^{-34}$ J s.
So the value of the Planck constant is 6.6×10^{-34} J s.

9.5 Applications of the photoelectric effect

Various devices based on the photoelectric effect have been developed. The photoelectric effect has practical applications in two key areas:
- the detection of light and the measurement of its intensity;
- the generation of electrical energy from light energy.

Photoemissive devices

These are based on the same principle as the apparatus that was originally used to investigate the photoelectric effect – it is shown in Fig. 9.7.

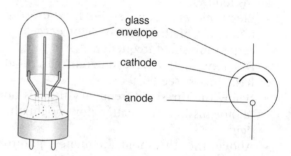

Fig. 9.7 A photoemissive cell

Light incident on the cathode causes electrons to be emitted. These move to the anode causing a current to flow in the external circuit. The current that flows is proportional to the incident light intensity.

Fig. 9.8 shows how a photocell is used to generate a signal from the sound track recorded on a movie film. The signal is recorded as variations in the transparency of the track. The changes in the intensity of the light transmitted by the track are converted to an electric signal by the photocell as the film passes. The signal is amplified and used to drive loudspeakers.

Fig. 9.9 shows how a photocell can be used to count boxes on a conveyor belt as they pass. As a box passes between the light source and the cell, the output of the cell decreases and then increases again. An external circuit counts the pulses electronically.

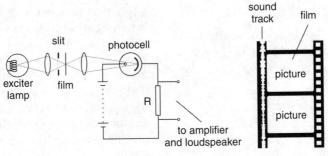

Fig. 9.8 Reading a movie sound track

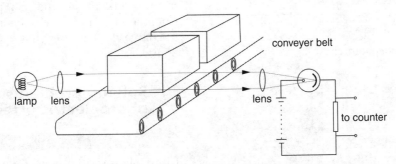

Fig. 9.9 Electronic counting

Photoconductive devices

Photoconductive devices change their electrical resistance when light falls on them. They are based on semiconducting materials, such as cadmium sulphide. Photons of visible light carry sufficient energy to free electrons from individual atoms in these materials. These electrons can then move freely through the material (like the free electrons in a metal) decreasing the electrical resistance.

Fig. 9.10(a) shows a light-dependent resistor (LDR) based on this principle. In the dark the resistance of the device is 10 MΩ. In daylight its resistance falls to 1 kΩ. Fig. 9.10(b) shows how an LDR may be used with a transistor switch to turn a light bulb on when it is dark and off when it is light.

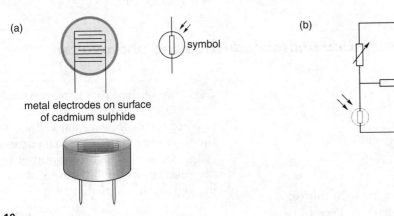

Fig. 9.10
(a) A light-dependent resistor (LDR)
(b) A light-sensing circuit

Photovoltaic devices

Photovoltaic devices generate an emf when light falls on them. They are based on the principle that a potential difference is produced at a junction between two dissimilar materials. When light is incident on the junction region, electrons are freed from atoms on either side. The potential difference causes the electrons to move in one direction across the junction (from negative to positive) and a current is thus generated in the external circuit. The selenium photocell shown in Fig. 9.11 is used as the light sensor in a photographic exposure meter.

The solar cells in Fig. 9.12 are constructed from layers of p- and n-type silicon (see Chapter 11). They generate sufficient current to charge the batteries that operate the traffic lights.

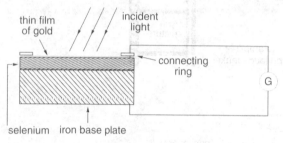

Fig. 9.11 Selenium photocell

Fig. 9.12 Solar cells

9.6 Solving problems on the photoelectric effect

Example

Ultraviolet radiation with frequency 1.2×10^{15} Hz is used in an investigation of the photoelectric effect with the metals listed in Table 9.1.
(a) Calculate the energy of the ultraviolet photons in joules and in electron volts.
(b) Which of the metals in the table will not exhibit photoelectric emission with radiation of this frequency?
(c) Calculate (i) the maximum kinetic energy, (ii) the maximum speed of the photoelectrons emitted when the radiation is incident on lead.
($h = 6.63 \times 10^{-34}$ J s; $m_e = 9.11 \times 10^{-31}$ kg; $e = 1.602 \times 10^{-19}$ C)

Activity 9.2 Investigating photoconductive and photovoltaic devices

You will need:
- a light-dependent resistor
- solar cells
- a laboratory multimeter
- a light source.

Method
1. Place the LDR under the light source. Connect the multimeter and set it to a resistance range. Observe how the resistance of the LDR changes when the light source is turned on and off.
2. Replace the LDR by a solar cell. Connect the multimeter and set it to a current range. Observe the current output from the solar cell as the brightness of the incident light is varied.

Table 9.1 Work functions of various metals

Metal	Work function, W (eV)
Chromium	4.44
Iron	4.60
Lead	4.01
Nickel	5.15
Platinum	5.63
Silver	4.44
Tin	4.28
Zinc	4.33

Answer

(a) $E = hf = 6.63 \times 10^{-34} \times 1.2 \times 10^{15}$
   $= 7.96 \times 10^{-19}$ J.

   Now, 1 eV $= 1.602 \times 10^{-19}$ J so
   7.96×10^{-19} J $= \dfrac{7.96 \times 10^{-19}}{1.602 \times 10^{-19}}$ eV
   $= 4.99$ eV.

(b) Nickel and platinum will not emit photoelectrons since their work functions are greater than the energy of the incident photons.

(c) *(i)* The Einstein equation gives
   $hf = hf_0 + \frac{1}{2}mv^2$.

   Therefore $KE_{max} = \frac{1}{2}mv^2 = hf - hf_0$.

   But $hf_0 = W$ so $KE_{max} = hf - W$
   $= 4.99 - 4.01 = 0.98$ eV.

   0.98 eV $= 0.98 \times 1.602 \times 10^{-19}$ J
   $= 1.57 \times 10^{-19}$ J.

   (ii) For the maximum speed,
   $v = \sqrt{\dfrac{2 \times KE}{m}} = \sqrt{\dfrac{2 \times 1.57 \times 10^{-19}}{9.11 \times 10^{-31}}}$
   $= 5.87 \times 10^5$ m s^{-1}.

Exercise 9.1

($h = 6.63 \times 10^{-34}$ J s; $m_e = 9.11 \times 10^{-31}$ kg; $e = 1.602 \times 10^{-19}$ C)

1. Referring to Table 9.1, calculate the threshold frequencies for photoelectric emission by tin and platinum.

2. The minimum frequency of light that will cause photoelectric emission from a certain metal surface is 5.25×10^{14} Hz. When the surface is illuminated by another type of radiation, electrons are emitted with a maximum speed of 7.25×10^5 m s^{-1}. Calculate:

   (a) the work function of the metal surface,
   (b) the maximum kinetic energy of the emitted electrons,
   (c) the frequency of the second source.

3. In an experiment to investigate the photoelectric effect, the stopping voltage, V_s, required to just prevent photoelectric emission is measured as a function of the frequency, f, of the incident light for lead, zinc and platinum.

   (a) Referring to Table 9.1, sketch graphs (on the same set of axes) of V_s against f for these metals.
   (b) Explain why there is a threshold frequency below which no photoelectric emission occurs. Is this frequency the same for all three metals? Explain your answer.
   (c) Name the constant that can be calculated from the gradient of the lines you draw.

9.7 Project to construct a burglar alarm

Use a photocell or an LDR to design a burglar alarm. Your design could be based on the circuit shown in Fig 9.10(b).

Construct your alarm so that a warning light or buzzer is triggered when a burglar passes between a light source and the light sensor.

Summary

- The emission of electrons by a metal exposed to light is called the photoelectric effect.
- Electrons are only emitted from the metal when the frequency of the incident light exceeds a certain threshold frequency.
- The threshold frequency is different for different metals.
- Below the threshold frequency there is no photoemission no matter how bright the light.
- Above the threshold frequency photoemission takes place no matter how dim the light.
- Above the threshold frequency the energy of the emitted electrons increases with increasing frequency of the light.
- The energy of the electrons does not depend on the brightness (intensity) of the light – increasing the intensity increases the number of electrons emitted, not their energy.
- Observations of the photoelectric effect may be explained by using the quantum theory of electromagnetic radiation – light travels as photons with energy hf, where h is the Planck constant and f is the frequency.
- The Einstein equation for the photoelectric effect is $hf = hf_0 + \frac{1}{2}mv^2$, where hf is the energy of the incident photons, f_0 is the threshold frequency, and $\frac{1}{2}mv^2$ is the kinetic energy of the photoelectrons emitted.
- The electron volt is defined as the energy change when one electron moves through a potential difference of 1 volt:
  $1\text{ eV} = 1.602 \times 10^{-19}\text{ C} \times 1\text{ V}$
  $= 1.602 \times 10^{-19}\text{ J}$.
- Photocells based on the photoelectric effect are used for the detection of light and the measurement of its intensity, and the generation of electrical energy from light energy.

Review questions

1. **(a)** Explain what is meant by the 'photoelectric effect'.
   (b) State the factors that affect photoelectric emission.
   (c) Explain the terms:
   (i) photon,
   (ii) threshold frequency,
   (iii) work function,
   (iv) electron volt.

2. **(a)** Calculate the energies, in eV, of photons of wavelengths *(i)* 5.0×10^{-7} m *(ii)* 1.0×10^{-8} m.
   (b) The work function for a given surface is 2.9 eV. Find *(i)* the threshold wavelength and *(ii)* the threshold frequency.
   ($h = 6.63 \times 10^{-34}$ J s, $c = 3.0 \times 10^8$ m s^{-1}, $e = 1.6 \times 10^{-19}$ C)

3. The threshold wavelength for a certain metal is 3.6×10^{-7} m. Calculate:
   (a) the threshold frequency,
   (b) the work function for the metal,
   (c) the voltage needed to stop the emission stimulated by light of wavelength 2.2×10^{-7} m.

4. The threshold frequency for photoelectric emission from sodium is 4.41×10^{14} Hz. Calculate:
   (a) the work function in electron volts,
   (b) the stopping potential for photoelectrons released by ultraviolet light of wavelength 3.00×10^{-7} m

5. The table below shows the stopping potential, V_s, and the corresponding frequencies for a certain photocell.

Stopping potential, V_s (V)	1.25	2.50	3.75	5.00	6.25
Frequency, ($\times 10^{15}$ Hz)	1.38	1.68	1.98	2.30	2.60

   (a) Plot a graph of stopping potential against frequency.
   (b) From your graph determine:
   (i) the threshold frequency,
   (ii) the threshold wavelength,
   (iii) the work function of the metal,
   (iv) a value for the Planck constant.

10 Radioactivity

Learning objectives

After completing the work in this chapter you will be able to:

1. define radioactive decay and half-life
2. describe the three types of radiations emitted in natural radioactivity
3. explain the detection of radioactive emissions
4. define nuclear fission and fusion
5. write balanced nuclear equations
6. explain the dangers of radioactive emissions
7. state the applications of radioactivity
8. solve numerical problems involving half-life.

10.1 Radioactive decay

In 1896 Henri Becquerel accidentally found that when he placed a uranium–potassium sulphate crystal on a photographic plate wrapped in thick black paper, the developed plate showed a blackened image of the crystal. No light had reached the plate, so why was it affected as though it had been in sunlight? Becquerel suggested that rays, which could pass through the wrapping paper, were given off from the uranium salt. Becquerel later discovered that similar rays were emitted from other uranium compounds. He concluded that uranium was responsible for the emission of the rays and this property was called **radioactivity**. He called the uranium a 'radioactive material'.

It was then found that radioactive materials caused the air around them to be electrically conducting. In 1898 Marie Curie and her husband, Pierre (who had invented an electrometer which could measure currents as small as 10^{-11} A), discovered two other radioactive elements, which were named polonium (after Poland, Marie's home country) and radium. In 1903 Becquerel and the Curies received the Nobel prize in physics for their work on radioactivity. It was soon discovered that radiation had medical uses but, sadly, it was not until much later that the dangers of such radiation were realised.

We now know that all naturally occurring elements with atomic numbers greater than 83 are radioactive. In addition, some of the elements with atomic numbers less than 83 have naturally occurring radioactive isotopes. Carbon is an example – naturally occurring carbon is made up of mainly carbon-12 (^{12}C), which is not radioactive, but also contains about 1% of carbon-14 (^{14}C), which is radioactive.

The nucleus

Fig. 10.1 shows the modern idea of the structure of an atom. Most of the mass of the atom is concentrated in the **nucleus** at the

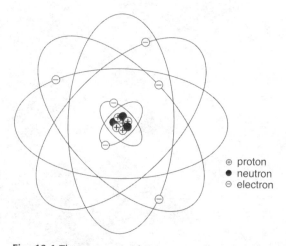

Fig. 10.1 The structure of the atom

centre. The nucleus contains **protons** and **neutrons**. Protons and neutrons are described as **nucleons**. The number of protons is the atomic number of the atom, Z. The number of protons plus the number of neutrons is the mass number, A.

The nucleus is orbited by **electrons**. In a neutral atom the number of electrons is equal to the number of protons.

The composition of a particular atomic nucleus is represented by $^A_Z X$, where X is the symbol for the element, Z is the atomic number and A the mass number. For example, $^{12}_6 C$, $^{238}_{92} U$ and $^{197}_{79} Au$.

During a *chemical* reaction there is no change in the *nuclei* of the atoms taking part. The atoms rearrange to form new substances because they bond together in different patterns, but nothing happens to the atomic nuclei. This is because chemical changes involve changes in only the arrangement of the electrons which form the outside part of atoms.

A *nuclear* reaction involves changes in the arrangement and/or energy of the protons and neutrons in atomic nuclei. **Radioactive decay** is a nuclear reaction in which an unstable nucleus loses energy by emitting particles and/or electromagnetic radiation.

Isotopes and nuclides

A **nuclide** is an atomic nucleus with a specific number of protons and neutrons – $^{12}_6 C$, $^{238}_{92} U$ and $^{197}_{79} Au$ are different nuclides. All nuclei may therefore be referred to as nuclides. **Isotopes** are atoms of the same element which have different mass numbers because they have different numbers of neutrons in their nuclei. The element carbon, for example, has three naturally occurring isotopes $^{12}_6 C$, $^{13}_6 C$ and $^{14}_6 C$.

Some isotopes are stable, but others are radioactive – for example ^{12}C is a stable nucleus but ^{14}C is radioactive, which means that it undergoes radioactive decay (sometimes called just 'decay').

Characteristics of radioactive substances

The atoms of radioactive elements are continually decaying (breaking down) into simpler atoms as a result of emitting radiation from the nuclei (see Fig. 10.2). Radioactive decay is a random, chance process. The probability of the decay of a nuclide of a particular type in a given time interval is constant, but it is not possible to predict exactly *when* a particular nucleus will decay. For a large number of nuclides of the same type, the fraction that decays in a given time interval is constant.

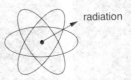

Fig. 10.2 Radioactive decay

The radiations from radioactive materials produce bright flashes of light when they strike certain compounds – the compounds are said to fluoresce. For example, rays from radium cause zinc sulphide to glow in the dark. A mixture of radium and zinc sulphide was once used to make the luminous paint that made the hands of watches glow in the dark.

The radiations released by radioactive substances cause ionisation of air molecules. Their energy knocks out electrons from the molecules, leaving them with a positive charge. Radiations from radioactive elements can penetrate the heavy black wrapping around a photographic film. When the film is developed, it appears black where the radiations struck the film.

Radiations from radioactive substances can destroy the germinating power of plant seeds, kill bacteria and burn or kill animals and plants. Radiation can also be used to kill cancers.

10.2 Half-life

Activity 10.1 Modelling radioactive decay

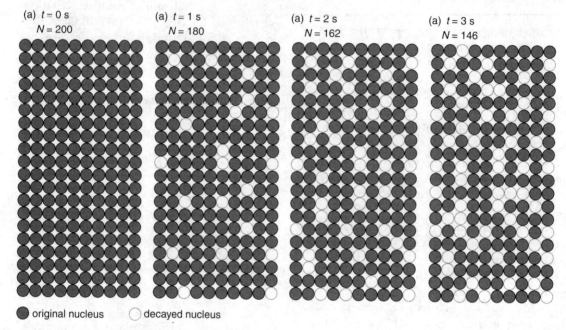

● original nucleus   ○ decayed nucleus

Fig. 10.3 A sample of 200 radioactive nuclei

Fig. 10.3(a) shows 200 radioactive nuclei at time $t = 0$ s. Suppose that the probability of a decay for this nuclide is 0.1 per second. This means that, on average, one-tenth of the nuclei will decay in the next second. We cannot say which ones these will be but we can predict that, since that are 200 atoms in the sample, 20 (from 0.1×200) will decay in the following second.

Note: In practice, since radioactive decay is a random process, the actual number of atoms that decay may be more or less than 20; but on average the sample will follow the decay law that we use to make predictions.

Fig. 10.3(b) shows the situation at time $t = 1$ s. There are now 180 radioactive atoms remaining. In the following second, one-tenth of these (in other words 18) will decay on average, leaving 162 at $t = 2$ s. At $t = 3$ s there will be $162 - (0.1 \times 162) = 146$ remaining. And so on …

Method
1. Copy Table 10.1 and complete it, rounding N to the nearest whole number at each stage.

2. Plot a graph of N against t.

Table 10.1 Radioactive decay with decay probability 0.1 per second

t (s)	N (number remaining)
0	200
1	180
2	162
3	
4	
5	
6	
8	
9	

(continued)

Activity 10.1 (continued)

3. From your graph determine the time taken for the number of nuclides to decrease:
   (a) from 200 to 100,
   (b) from 100 to 50,
   (c) from 50 to 25.

Activity 10.1 demonstrates that the rate of radioactive decay (the number of decays per second) is proportional to the number of nuclei present. If the number of nuclei present is N then we can write rate of decay $\propto N$, or rate of decay $= \lambda N$ where λ is the probability per second of a single nuclide undergoing decay. λ is called the **decay constant**. Each type of radioactive nuclide has a different value for λ.

The shape of the graph of N against t is an example of exponential decay. The magnitude of its gradient at any instant is the rate of decay (the number of decays per second) at that instant. This decreases as the number of radioactive nuclides decreases.

A special feature of exponential decay is that the time taken for the number of nuclides to decrease by half (from 200 to 100, or 100 to 50 for example) is a constant. This is called the **half-life** for the decay and is given the symbol $t_{\frac{1}{2}}$.

Example

What is the half-life of the sample in Activity 10.1?

Answer

Look at your response to Step 3 in Activity 10.1.

The half-life of a nuclide is characteristic of that nuclide – it is governed by the structure of the nucleus. It is not affected by temperature, pressure or any physical condition. Half-lives vary from many years to a fraction of a second. Some half-lives are shown in Table 10.2

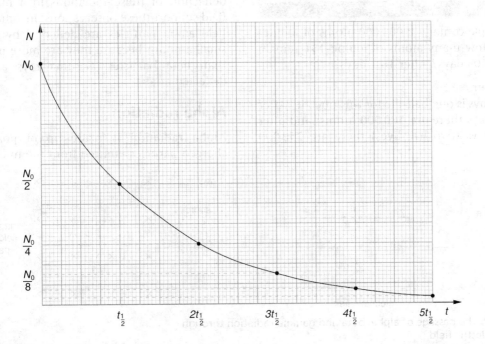

Fig. 10.4 The rate of radioactive decay

Table 10.2 Half-lives of some nuclides

Nuclide	Half-life	Notes
H-3	12.26 y	Hydrogen isotope called tritium
C-14	5730 y	Found in the atmosphere
N-12	0.01 s	Very hard to study
K-40	1.26×10^9 y	Found in the human body
Co-60	5.24 y	A common γ source
I-131	8.05 d	Used in medicine
Cs-137	30.0 y	Spread by the Chernobyl explosion
Sr-90	27.7 y	A common laboratory β source
Rn-220	54 s	Source of lung cancer
U-238	4.5×10^9 y	The common isotope of uranium
Pu-240	6580 y	Used in nuclear weapons

Example

A sample contains 1 000 000 atoms of iodine-131. How many atoms of iodine-131 remain after 8.05 days? After 16.1 days?

Answer

8.05 days is one half-life for this nuclide. After 8.05 days there are 500 000 atoms; and after 16.1 days (two half-lives) there are 250 000 atoms.

10.3 Types of radiations and their properties

In 1897 Ernest Rutherford showed that the invisible radiation discovered by Becquerel had at least two components with different properties. Rutherford called them **alpha** (α) and **beta** (β) radiation. A third component of this radiation was later discovered by Pierre Curie – it was called **gamma** (γ) radiation. The properties of gamma radiation differ from those of alpha and beta radiations. However, all three originate from the nucleus of the atom.

Most radioactive substances emit at least two kinds of radiation simultaneously (usually alpha and gamma, or beta and gamma). When a radioactive source that emits all three kinds of radioactivity is directed into an electric field, the three radiations behave differently, as shown in Fig. 10.5.

Alpha radiation is attracted towards the negative plate of the electric field. Beta radiation is attracted towards the positive plate. Gamma radiation is unaffected and passes straight through. This shows that alpha radiation carries a positive charge, beta radiation has a negative charge, while gamma radiation has no charge associated with it. The deflection of these radiations in a magnetic field support these observations. In a magnetic field alpha particles are deflected by only a small amount because they are more massive than beta particles. Gamma rays are not deflected at all.

Alpha radiation

Alpha radiation is made up of positively charged particles emitted from the nucleus of

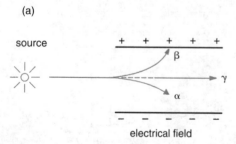
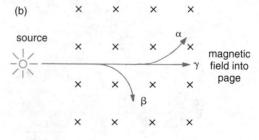

Fig. 10.5 The passage of alpha, beta and gamma radiation through
(a) an electric field
(b) a magnetic field

radioactive atoms. Measurement of their charge to mass ratio, made by J.J. Thomson at the beginning of this century, suggested that they were helium nuclei, and this was confirmed by Rutherford and others. In nuclei, *two* is a number that gives stability – nuclei with *two* of anything are very stable. The helium-4 nucleus ($^4_2$He), which contains two protons *and* two neutrons, is particularly stable. Since it contains two protons, the alpha particle carries a charge of +2 (where the unit of charge is the charge on a single electron or proton).

Alpha particles travel at about 5% of the speed of light. Because of their +2 charge, they tend to attract electrons away from nearby atoms and so cause ionisation in a gas. Alpha particles have a low power of penetration, travelling only a few centimetres in air. They can be stopped by a thin sheet of paper or the outer layer of skin. Because they cannot travel very far, even in a gas, all their ionising power becomes concentrated in a small volume. Their effect is therefore intense where it occurs.

Alpha particles can be detected by a photographic plate, in a cloud chamber (by straight, thick, short tracks all of about the same length) and by a spark counter. The most energetic alpha particles can be detected by a Geiger–Muller (GM) counter. Americium-241 is a good laboratory source of alpha radiation.

Beta radiation

Beta radiation is made up of particles emitted from the nuclei of radioactive atoms. Measurement of their charge to mass ratio shows that beta particles are fast-moving electrons, like cathode rays. They are much lighter than alpha particles and have a charge of −1. Their speed ranges from 30% to 99% of the speed of light. They are much less ionising than alpha particles but are far more penetrating. The most energetic beta particles have a range in air of a few metres, and are stopped by about 5 mm of aluminium or 1 mm of lead.

The term 'beta particle' is reserved for an electron which comes from the *nucleus* of an atom. The electrons outside the nucleus are simply called electrons. Because electrons have a dual nature acting as particles and as waves, we can speak of either 'beta particles' or 'beta radiation'.

Beta particles can be detected by a photographic plate, in a cloud chamber (by thin and twisted tracks) and by a GM counter. Strontium-90 is a good laboratory source of beta radiation.

Gamma radiation

Gamma radiation is high-energy electromagnetic radiation emitted from the nucleus of an atom. It travels at the speed of light and carries no charge. The nature of gamma radiation was not established until 1914. Gamma emission generally occurs after alpha or beta emission. Its effect is to carry away excess energy from the excited nucleus.

Like X-rays, gamma rays are very penetrating, but their ionising power is very low. Their intensity can be reduced significantly by several centimetres of lead. As shown in Fig. 10.6, they have shorter wavelengths than X-rays. There is no sharp dividing line between ultraviolet rays and X-rays, or between X-rays and gamma radiation.

Gamma rays can be detected by a photographic plate, in a cloud chamber (by straight tracks spreading out from the gamma ray source) and by a GM counter. Cobalt-60 is a good laboratory source of gamma radiation.

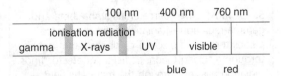

Fig. 10.6 Part of the electromagnetic spectrum

Table 10.3 summarises the properties of radioactive emissions and Fig. 10.7 compares their penetrating power with respect to human flesh.

Table 10.3 A comparison of alpha, beta and gamma radiation

Radiation type	Ionising power	Penetrating power	Range in air (m)	Electric charge	Absorbed by
Alpha (α)	High	Low	~ 0.05	+2	Paper
Beta (β)	Medium	Medium	~ 3	−1	~ 5 mm aluminium
Gamma (γ)	Low	High	~ 100	0	~ 3 cm lead halves intensity

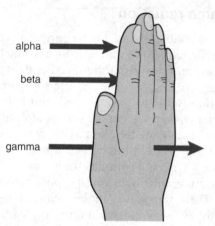

Fig. 10.7 The penetrating powers of radiation through the human body

Exercise 10.1

1. Is radioactivity the emission of rays or particles from **(a)** outer space, **(b)** atomic nuclei, **(c)** orbiting electrons or **(d)** X-rays?

2. Which type of radioactivity is:
(a) the most penetrating,
(b) the most massive,
(c) not composed of particles?

3. A radioactive source emits alpha, beta and gamma radiations. Radiation from the source passes through a piece of cardboard and then through a strong magnetic field. A detector finds radiation passing through the magnetic field as well as to one side of the field. Which types of radiation are being detected: **(a)** alpha only, **(b)** beta only, **(c)** gamma only or **(d)** beta and gamma?

4. Which of the following radiations are not deflected by an electric field: **(a)** alpha, **(b)** beta, **(c)** gamma, **(d)** visible light?

10.4 Detectors of radiation

Photographic plates

Photographic plates can be used to detect radioactive rays – such plates are blackened by the rays. Scientists working with radiation are required to wear a sealed badge containing a photographic film as a personal radiation monitor (see Fig 8.3). The film is replaced and developed at regular intervals and the amount of black-ening of the film indicates the level of exposure. If the level is too high, the worker must be checked by a doctor and the area of work checked for leakage of radiation. There is a limit as to how much radiation a worker can safely receive during a year.

Ionisation detectors

The three radiations we have studied (alpha, beta and gamma) all deposit energy in matter as they pass through it. This energy may eject electrons from atoms, producing ionisation. The ions formed may in turn be detected by the resulting changes in electrical conductivity, or the flashes of light emitted when electrons and ions recombine.

The detectors described below use ionisation in various ways to detect and measure radiation.

Spark counter

When nuclear radiations pass through a gas between a positively charged wire grid and a negatively charged metal plate, ionisation occurs – see Fig. 10.8. The ions and electrons produced enable a large current to suddenly pass through the air under the high voltage

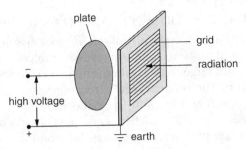

Fig.10.8 A spark counter

between the grid and the plate. A spark can be seen and heard, or it may be registered by an electronic device (such as a scaler).

Cloud chamber

A diffusion cloud chamber is used in laboratories to reveal the tracks of charged particles (α and β radiations) – see Fig. 10.9. The base of the chamber is cooled by dry ice to about −80°C. A felt ring inside the top of the chamber is moistened with alcohol. The alcohol vapour diffuses downwards, becoming cooled and ready to condense. Each time a particle is emitted from a radioactive source, it produces ions along its path and the alcohol vapour then condenses around these ions. The condensed alcohol droplets reflect light and so can be seen as narrow white lines (see Fig 10.10).

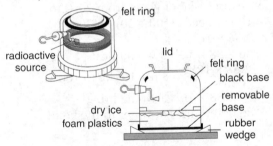

Fig. 10.9 A diffusion cloud chamber

The cloud chamber provides evidence that something is actually being emitted from radioactive materials. Cloud chamber photographs do not show the actual radiation, but only the alcohol droplets which form on the ions produced by such radiation.

Example

In Fig. 10.10 how many different alpha particles were entering the chamber? Explain your answer.

Fig. 10.10 Tracks of alpha particles in a cloud chamber

Answer

Two – the tracks are of two different lengths, showing particles with two different energies.

Geiger–Muller tube

The Geiger–Muller tube is one of the most important instruments for detecting radiations – one of these instruments is shown in Fig. 10.11. When nuclear radiation enters the GM tube, the ions which are produced allow a sudden large current pulse to pass through the tube. This pulse can be detected by a scaler or a ratemeter. A scaler records the total number of counts, whereas a ratemeter records the number of pulses or counts per second. GM tubes are not very efficient. They record about 10% of the beta radiation received and less than 1% of the gamma radiation.

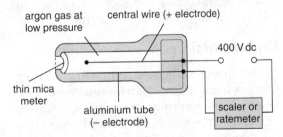

Fig. 10.11 A Geiger–Muller counter

The combination of a GM tube and a scaler or ratemeter is usually referred to as a Geiger–Muller counter. GM counters are simple and, with care, long lasting. They have the disadvantage that a given radiation either triggers the tube or it does not – the response does not depend on the energy of the radiation.

Scintillation counter

One of the early methods of detecting radioactivity was with the use of a fluorescent screen in conjunction with a microscope. When nuclear radiation strikes the screen, light spots – scintillations – are produced. The modern scintillation counter consists of a fluorescent crystal placed in contact with a photomultiplier. When radiation strikes the crystal, light is emitted. The light is detected by the photomultiplier and amplified by its electronic circuit. Such a counter is much more sensitive than a simple fluorescent screen. It has the advantage that the intensity of the light pulse which it produces depends on the energy deposited in the crystal.

10.5 Nuclear fission and nuclear fusion

The particles in the nuclei of atoms are held together by very strong forces. To break up a nucleus requires the input of huge amounts of energy. Similarly, when a nucleus is formed huge amounts of energy are released.

Einstein showed that mass and energy can be converted into each other. They are related by the equation $E = mc^2$, where E is the energy in joules, m is the mass in kilograms and c is the velocity of light. The square of the velocity of light is a huge number, so small masses can give rise to huge amounts of energy.

Nuclear fission

Nuclear fission is the process in which a heavy nucleus splits into nuclei with smaller masses. One or more neutrons may also be released. The fragments weigh very slightly less than the original nucleus. The lost mass appears as energy, so a large amount of energy is released. The energy is in the form of gamma rays and the kinetic energy of the fragments, which is ultimately converted into heat energy by collisions with other atoms.

One way in which a uranium-235 nucleus can undergo fission is given by the equation:

$$^{235}_{92}U + ^{1}_{0}n \rightarrow ^{138}_{56}Ba + ^{95}_{36}Kr + 3^{1}_{0}n + (4.8 \times 10^{-11} \text{ J})$$

This is an example of an *induced* fission reaction – the uranium nucleus is induced to split (undergo fission) when a neutron collides with it. The products of the reaction, isotopes of barium and krypton, are radioactive.

Many similar reactions are possible each resulting in different products, for example

$$^{235}_{92}U + ^{1}_{0}n \rightarrow ^{134}_{54}Xe + ^{100}_{38}Sr + 2^{1}_{0}n + \text{energy}$$

Example

One mole (a small handful) of uranium contains 6×10^{23} atoms.
(a) If all these atoms undergo fission in the way described above, how much energy is released?
(b) How much water, at room temperature, could be heated to 100 °C by this amount of energy?

Answer

(a) Fission of one atom of uranium-235 releases 4.8×10^{-11} J.
Fission of one mole of atoms releases $4.8 \times 10^{-11} \times 6 \times 10^{23} = 2.88 \times 10^{13}$ J.
(b) Raising 1 kg water through 1°C requires 4200 J. If room temperature is 20°C then heating 1 kg of water from 20°C to 100°C needs $(100 - 20) \times 4200 = 3.36 \times 10^5$ J.
So 2.88×10^{13} J would heat
$\dfrac{2.88 \times 10^{13}}{3.36 \times 10^5} = 8.6 \times 10^7$ kg to 100°C.
So 2.88×10^{13} J would bring 86 000 tonnes of water to the boil!!

The neutrons released by the fission of a nucleus can trigger off the fission of other nuclei, which in turn release more neutrons as shown in Fig. 10.12. In this way, a chain reaction occurs releasing an enormous amount of energy in a short time. For a chain reaction to occur, the mass of the uranium-235 sample has to exceed a certain value, called the critical mass.

The neutrons emitted during fission are usually moving too fast to be absorbed by other fissionable nuclei. To be effective, the neutrons must be slowed down by a suitable

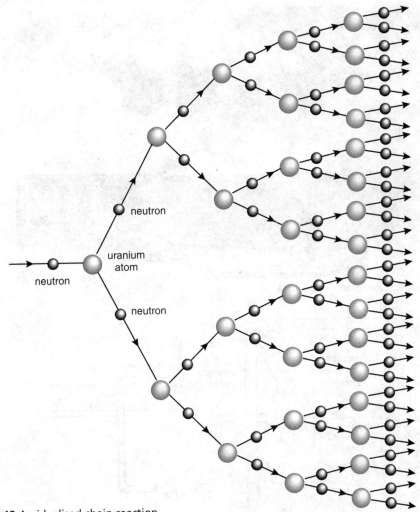

Fig. 10.12 An idealised chain reaction

material called a moderator – for example, graphite or water made from the isotope of hydrogen containing a proton and a neutron (deuterium, ^2_1H). Slowing down the neutrons ensures that more collisions occur with the remaining uranium nuclei. The first successful fission reaction was carried out by Hahn and Strassman in 1939. It was later discovered that the element plutonium also undergoes fission and produces more neutrons when bombarded with slow neutrons.

Almost pure uranium-235, or plutonium-239, is used in an atomic bomb – see Fig. 10.13. The destructive power of the atomic bomb comes from the energy released in a chain reaction. The first atomic bomb was exploded in 1945 at a test site in New Mexico, USA. Shortly afterwards, atomic bombs were dropped on Hiroshima and Nagasaki in Japan causing the destruction of the two cities. The radiations and radioactive fallout from the bomb harmed many living things for years afterwards. The dropping of these bombs ended World War II.

Nuclear energy

The same chain reaction system can be used in a controlled way to liberate useful energy from fission. The world's first commercial nuclear power station was opened in 1956 at Calder Hall, England. In a nuclear power station, the fuel is arranged so that it can never explode as an atomic bomb does. The supply of neutrons is controlled so that the reaction is self-sustaining but no more. Fission in a nuclear reactor provides energy to

Chapter 10: Radioactivity   113

Fig. 10.13 An American test explosion at Bikini Atoll

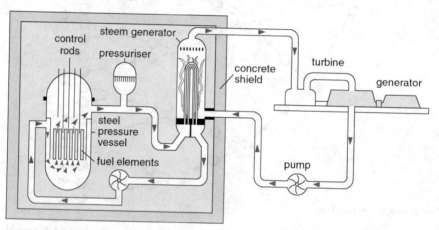

Fig. 10.14 A simplified pressurised water reactor

convert water into steam. The steam is used to turn a turbine, which is connected to a generator – the generator produces electricity. Careful control of the chain reaction ensures that energy is released at a steady rate.

There are many different types of nuclear reactor. The pressurised water reactor (see Fig. 10.14) is the most common type of reactor used today. The fuel (uranium oxide enriched with uranium-235) is contained in a steel pressure vessel. Energy released during the fission process heats the water to around 325°C. The heated water is carried in pipes at high pressure (155 atmospheres) to the steam generator. The pressurised water also acts as a moderator to slow down the fission neutrons, thereby maintaining the chain reaction. Boron–steel control rods inserted in the fuel elements are used to control the reaction. The boron absorbs the neutrons which are capable of causing further fission. The chain reaction, and hence the temperature, can be controlled by raising or lowering the control rods. In an emergency all the control rods can be lowered quickly to shut down the reactor.

The whole reactor is enclosed in a concrete shield to prevent any nuclear radiation from escaping and to protect workers. Used fuel elements are sent to a reprocessing plant when their activity drops below a certain level. At the reprocessing plant, unused uranium is separated from the radioactive waste. Unused uranium can be reused but the waste has to be carefully stored, as it can remain radioactive for thousands of years. Low-activity waste material is dumped at sea in steel and concrete drums. Long-life, highly active waste is stored in high-integrity tanks, awaiting safe disposal.

The first man-made nuclear reactor was made and operated by Enrico Fermi and his colleagues in 1942. But about 2 billion years ago, deposits of uranium in some parts of Africa had enough of the neutron-emitting uranium-235 to become natural chain reactors.

The water content of a rich vein of uranium ore at Oklo, in Gabon, was sufficient to moderate naturally emitted neutrons so that they took part in a chain reaction. It is thought that this 'reactor' operated, on and off, for thousands of years, producing energy at kilowatt levels.

Nuclear accidents

Over the years there have been a number of accidents at nuclear power stations which have released radioactive pollutants. The most serious was in 1986 at Chernobyl, in what was then the USSR. During a test, the water which cooled the reactor boiled so quickly that the cooling system exploded. The force was such that the 1000 tonne concrete lid of the reactor was ripped off. Massive amounts of radionuclides from the reactor core were thrown into the air and in time were detected all over the northern hemisphere. This was not a nuclear explosion, but the polluting effects were severe.

Nuclear fusion

In nuclear fission, a heavy nucleus is split into two to release energy. The reverse process can also produce large amounts of energy. The combination of two light nuclei to form a heavier nucleus is called **nuclear fusion**. See Fig. 10.15 for an example – two nuclei of hydrogen-2 (deuterium) can be smashed together to form a nucleus of helium-3 and a

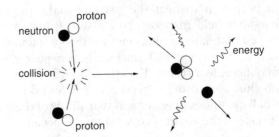

Fig. 10.15 Nuclear fusion

neutron. The reaction can be written as

$$^2_1H + {}^2_1H \rightarrow {}^3_2He + {}^1_0n + (4.8 \times 10^{-13}\ J)$$

The total energy released in this reaction is less than that of uranium fission, but the energy per nucleon is much greater. In other words, less material is required to produce the energy. Nuclear fusion is the energy source for the Sun and stars. The solar energy received by us on Earth is due to the uncontrolled fusion of hydrogen nuclei in the Sun. Uncontrolled fusion here on Earth can be seen in the hydrogen bomb. The initial high temperature required is obtained by using a fission bomb to trigger the fusion. A hydrogen bomb releases much more energy than an atomic bomb. Scientists around the world are attempting to produce controlled fusion in the laboratory. Controlled fusion could be an almost limitless source of energy for the future.

To start the fusion reaction of two nuclei they must be brought sufficiently close to each other. It is not easy to do this as their positive charges repel each other with a very large electrical force. One way to bring the nuclei together is to heat them up to an extremely high temperature (10^8 K) so that they gain enough kinetic energy to overcome the repulsion. At these temperatures, however, containment is difficult because any material container would be vaporised. One solution is to make the particles circulate in an intense magnetic field. The particles are charged, so they have a magnetic field of their own and with the right design the hot material never touches the containing walls.

The problem with nuclear fusion as a source of energy is that, at present, more energy has to be put in to heat the gas and make the magnetic field than can be got out. But it is hoped that, later in this century, this problem will be overcome and nuclear fusion power may become a reality. If so then there will be no shortage of fuel – hydrogen-2 is found in small quantities in water and two-thirds of the surface of the Earth is covered by the oceans.

10.6 Nuclear equations

In nuclear changes, just as in chemical changes, *nothing* is ever lost or gained. For example, in some reactions a neutron transforms into a proton by emitting an electron, but the total number of nucleons (protons plus neutrons) and the total charge remain the same. These two rules can be used to balance nuclear reaction equations:
- the total of the mass numbers on the left-hand side must equal the total of the mass numbers on the right-hand side;
- the total of the atomic numbers on the left-hand side must equal the total of the atomic numbers on the right-hand side.

Example 1

An atom of carbon-14 is formed in the Earth's upper atmosphere when a neutron bombards a nitrogen atom. What other product is formed in this nuclear reaction?

Answer

$^{14}_{7}N + ^{1}_{0}n \rightarrow ^{14}_{6}C + ^{x}_{y}?$

The total mass number on the left-hand side = 14 + 1 = 15.
So on the right-hand side, 14 + x = 15, so x = 1.
The total atomic number on the left-hand side = 7 + 0 = 7.
So on the right-hand side, 6 + y = 7, so y = 1.
The unknown particle on the right-hand side has atomic number 1 (y), and mass number 1 (x), so it must be an atom of hydrogen, $^{1}_{1}H$.

$^{14}_{7}N + ^{1}_{0}n \rightarrow ^{14}_{6}C + ^{1}_{1}H$

This checking system can be used to find out whether a nuclear reaction is balanced (which does *not* necessarily mean that it is possible) or to discover any one missing number.

Example 2

What other product is formed when polonium-218 decays to astatine?

$^{218}_{84}Po \rightarrow ^{218}_{85}At + ^{x}_{y}?$

Answer

The total mass number on the right-hand side has to be 218. So 218 + x = 218, so x = 0. The unknown particle has a mass number of 0.

The total atomic number on the right-hand side has to be 84.

So 85 + y = 84, so y = –1.

No nucleus has an atomic number of –1 but if we think of this number as 'the charge on the nucleus' then it is clear that this refers to an electron (equivalent to a beta particle). The unknown particle is an electron, $^{0}_{-1}e$.

The equation shows that an atom of polonium-218 decays by emitting an electron (a beta particle) and forms an atom of astatine-218.

Effects on the nucleus

Alpha decay

Fig.10.16 shows a nucleus forming another nucleus and an alpha particle – this is what happens in alpha decay. The atomic number, Z, of the nucleus goes down by two, and the mass number, A, goes down by four.

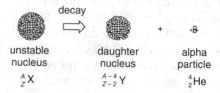

Fig. 10.16 Alpha decay

An example of alpha decay is the decay of uranium-238, producing thorium-234.

$^{238}_{92}U \rightarrow ^{234}_{90}Th + ^{4}_{2}\alpha$

[or $^{238}_{92}U \rightarrow ^{234}_{90}Th + ^{4}_{2}He$]

We can show this change in another way – by drawing a grid with mass number, A, vertically and atomic number, Z, horizontally, as shown in Fig. 10.17. The original nucleus loses four units of mass and two units of positive charge.

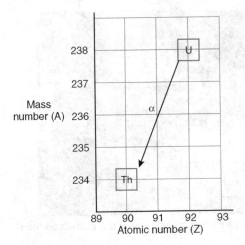

Fig. 10.17 Alpha decay of uranium-238 to thorium-234

Beta decay

A beta particle is an electron, so in beta decay electrons are emitted from nuclei. Electrons do not exist inside the nucleus but can be produced if a neutron changes into a proton.

$${}^{1}_{0}n \rightarrow {}^{1}_{1}p + {}^{0}_{-1}e$$

The atomic number, Z, is increased by one (because a new proton has been created) but the mass number, A, remains unchanged. For example, when strontium-90 undergoes beta decay, the daughter nucleus becomes yttrium-90.

$${}^{90}_{38}Sr \rightarrow {}^{90}_{39}Y + {}^{0}_{-1}e$$

Showing this change on a grid gives the diagram in Fig. 10.18.

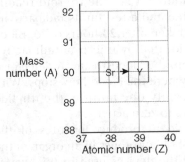

Fig. 10.18 Beta decay of strontium-90 to yttrium-90

Gamma emission

Gamma rays are electromagnetic radiations, not material particles. When a nucleus emits a gamma ray the nucleus keeps the same atomic number, Z, and the same mass number, A. The gamma ray only carries away energy and the nucleus becomes more stable.

Gamma emission may occur following an alpha or beta emission in which the product nuclide is left in an excited (raised energy) state. The excess energy is emitted as a photon of electromagnetic radiation. Cobalt-60 is a common gamma-emitting nuclide.

$${}^{60}_{27}Co^* \rightarrow {}^{60}_{27}Co + \gamma$$

(* = in excited, unstable state)

Exercise 10.2

1. When two atoms of deuterium, ${}^{2}_{1}H$, fuse together they form a neutron and one new atom. What is that atom?

2. A radioactive nuclide, of atomic number Z, emits a beta particle followed by a gamma ray. What is the atomic number of the new nuclide? Is it **(a)** $Z + 1$, **(b)** $Z - 2$, **(c)** Z or **(d)** $Z - 1$?

10.7 The hazards of radioactivity

Effects on living things

Nuclear radiation absorbed by the human body can cause severe damage at the molecular scale. Exposure to a single large dose of nuclear radiation, for example through proximity to a nuclear explosion, can produce radiation sickness leading to death within days or weeks. Chronic long-term exposure to lower levels of radiation, for example from contamination following a nuclear accident or the incorrect use of radiation sources in industry, increases the risk of cancers such as leukaemia and lung cancer, and also the incidence of birth defects in the children of those exposed.

We are continually exposed to radiations from outer space as well as from the Earth, the air, building materials, our food and, indeed, our own bodies.

The radiations coming from **outer space** are called cosmic rays. Cosmic radiation accounts for about 16% of our annual natural radiation dose. The Earth itself emits radiations because it contains various radioactive substances. This natural radiation from outer space and the Earth can be detected by a Geiger–Muller tube – it is called **background radiation**. If a GM counter is set up in a normal room it records a low count rate (about 30 counts per minute, though this varies from place to place). The amount of radiation to which we are exposed depends on where we live.

In some places, the **air** we breathe contains radioactive radon gas, which is naturally produced from the decay of the element uranium present in many rocks. This gas cannot be seen, tasted or smelled. The annual intake of radon is increased in modern, draughtproof houses. The United States Environmental Protection Agency estimates that inhaled radon probably contributes to 20 000 lung cancer deaths per year in America and kills 1 in 10 000 people per year world wide. In some parts of the UK the government provides money for testing air and for making houses safe against radon gas.

The **food** we eat or drink contains small amounts of radioactive materials. For example, many foods contain tiny amounts of radioactive potassium-40, lead-210 and/or polonium-210. These radioactive substances become concentrated in fish and other animal meat and are passed to us when we eat these animals as foods.

Precautions

It is important to realise that we have always lived with background radiation. The amount of background radiation varies greatly from place to place but no one should worry about visiting a region with a higher than average background radiation or taking a trip in an aeroplane. But it is sensible that the extra doses of radiation we get should be kept as small as possible. To keep doses small:
- minimise the time spent near any artificial radiation source;
- never eat or drink in the laboratory;
- keep an adequate distance from such sources;
- always use suitable shielding as instructed by trained personnel.

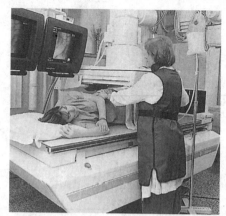

Fig. 10.19 Radiographer wearing a protective lead apron

10.8 The applications of radioactivity

The radiations emitted from radionuclides (radioactive nuclides) have many applications. Such uses are commonly based on:
- the penetrating power of the radiation; or
- its ability to kill living cells; or
- the energy carried by the radiation.

Medical uses

Radioisotopes can be used as **medical tracers**. A small amount of a radioisotope is put into a person's body. It can be traced through the body system because the radiation it emits can pass through body tissues. For example, the function of the thyroid gland in the neck can be studied by drinking a small amount of radioactive iodine-131. The thyroid readily takes up iodine and a test on its radioactivity can be undertaken after 24 hours or so. Blood clots can be located by injecting radioactive sodium-25 into the bloodstream. A detector is used to find where the blood flow stops. With a half-life of only one minute, the nuclide quickly decays to zero activity.

It is possible to produce pictures of the distribution of radioisotopes in an organ of the body. One of the devices used in this way is the gamma camera. It detects gamma rays from the organ and forms an image on a TV-type screen.

Another important medical application of radiation is the treatment of cancer. Gamma rays emitted from a strong source of cobalt-60 can be used to kill cancer cells.

As energy sources

A radioactive material can be used as the energy source for an artificial pacemaker for the heart, for example. The radioisotope-powered pacemaker supplies regular electrical signals to shock a defective heart into a normal beating cycle. The pacemaker is surgically implanted in the body.

Radioisotopes are also used in small nuclear electric power systems – such systems are used in space stations and remote stations on earth.

Gamma sterilisation

The preparation of medical supplies has made increasing use of radiations for sterilisation. Packages, like those containing syringes and needles, are irradiated by gamma rays and any germs present are killed. This eliminates the need for heavy and cumbersome sterilisation apparatus.

Radiation is also used in food preservation. For example, some freshly caught sea foods have an ice-storage life of about 15 days. After being exposed to gamma rays, they are still in good condition after seven weeks in ice.

Radiodating

Because radionuclides decay at known rates, and because some have very long half-lives, they can be used to pinpoint the age of ancient materials.

As living things grow they take in a fixed proportion of radioactive carbon-14 along with the non-active carbon-12, which is the basis of most living tissue. When the plant or animal dies it stops taking in carbon and its existing carbon-14 decays. The isotope is known to have a half-life of 5730 years and the change in its activity as it decays can be used to estimate how long ago the living thing died. The age of the material can be found if about 5 mg of material can be spared for analysis of carbon-14 activity, providing that it is less than about 20 000 years old.

Fig. 10.20 shows a photograph of a human body found buried in ice in northern Europe – it was found in 1991. Initially it was thought to be only 200 years old but radio-

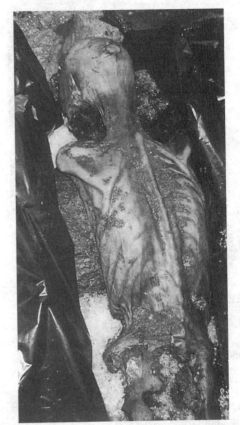

Fig. 10.20 A human body found buried in ice in northern Europe

carbon dating indicated that the man died about 5200 years ago.

Industrial uses

Radiation has many applications in industry. For example, the penetrating power of gamma rays is used to detect hidden flaws in metal castings; beta rays are used to measure and monitor the thickness of various flat objects – the amount of radiation absorbed by the object is related to its thickness; radioactive materials can be used as tracers to investigate and monitor the flow of liquids in pipelines.

In the textile industry, irradiation with beta rays fixes various chemicals onto cotton fibres. This can help to produce, for example, permanent creases in clothing such as trousers.

Agricultural uses

Radionuclides are used as tracers for studying plants, insects and animals. For example,

Fig. 10.21 Autoradiograph of pelargonium leaf

phosphorus-32 can be added to fertilisers and the phosphorus taken up by plants can be measured and its distribution investigated.

Radiation has been used to control the screw-worm fly pest in South America. A large number of the male of the species were exposed to gamma radiation. When the males were released back into the wild and mated with wild females, sterile eggs resulted and no new flies were born.

The points of photosynthesis in a leaf can be revealed by growing it in air containing carbon-14. The presence of this radioactive nuclide in the leaf is then revealed by putting the leaf onto a photographic plate and 'letting it take its own picture' – see Fig. 10.21.

10.9 Solving problems on radioactivity

Exercise 10.3

1. (a) An isotope of nitrogen can be represented as $^{14}_{7}N$. What is the significance of the numbers 14 and 7?

(b) Find a and b in the radioactive decay equation $^{214}_{82}X \rightarrow {}^{a}_{b}Y + {}^{0}_{-1}e + \gamma$.

(c) What are the values of p and q in the nuclear equation $^{p}_{q}X \rightarrow {}^{230}_{90}Y + 2\alpha$

2. A nuclide, F, has a half-life of 2.5 hours. What percentage of the original number of atoms of the isotope would be left after 10 hours?

3. (a) Copy and complete Fig. 10.22 to show how the particles and rays are deflected and at which material each of them is stopped.

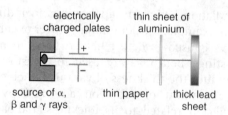

Fig. 10.22

(b) Fig. 10.23 shows the deflection of radiations from a radioactive source by a uniform magnetic field. State and explain two factors which bring about the differences in the deflections.

4. (a) Radon gas, $^{222}_{86}Rn$, decays by emission of one alpha particle. Find, by use of an equation, the product of the reaction.

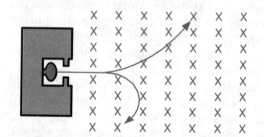

Fig 10.23

(b) Explain the terms *fusion* and *fission* in nuclear reactions.

(c) Explain how carbon-14 is formed in the atmosphere.

(d) Give two uses of cobalt-60 as a radioactive source.

5. (a) Explain what is meant by the *spontaneous nature* of radioactive decay.

(b) Explain what is meant by *half-life* and how the concept depends on the random nature of radioactive decay.

(c) A sample of a certain nuclide, which has a half-life of 1500 years, has an activity of 32 000 counts per hour at the present time.

(i) Plot a graph of the activity of this sample over the period in which it will reduce to one-sixteenth of its present value.

(ii) If the sample of the nuclide could be left for 2000 years, what would be its activity then?

Summary

- Changes in the nuclei of atoms can cause subatomic particles or electromagnetic waves to be emitted.
- Such emissions are called radioactivity.
- Radioactivity comes from naturally occurring elements and from artificially made nuclides.
- Common emissions are alpha (α), beta (β) and gamma (γ) radiation.
- Alpha particles are helium nuclei, $^4_2$He.
- Beta particles are electrons which have come from an atomic nucleus.
- Gamma rays are electromagnetic radiation.
- Alpha, beta and gamma radiations have different penetrating powers through matter.
- Nuclear equations can be written like chemical equations, by balancing the mass numbers and atomic numbers of particles on both sides of the equation.
- In alpha decay the mass number of the nucleus goes down by 4 and the atomic number goes down by 2.
- In beta decay the mass number of the nucleus does not change but its proton number goes up by 1.
- In gamma decay there is no change in either mass number or atomic number.
- The half-life of a nuclide is the time taken for half the atoms present to decay.
- The half-life of a nuclide is not affected by physical conditions.
- The number of radioactive atoms present in a sample decays exponentially.
- Radioactivity can be detected by a spark counter, a cloud chamber, a Geiger–Muller tube or a scintillation counter.
- Energy is released when heavy atomic nuclei split into smaller nuclei – the energy comes from the loss of mass in the fission process.
- Energy is also released when light nuclei fuse together to form heavier nuclei.
- Energy released in these ways can be used in weapons or for peaceful purposes.
- Radioactivity is used as an energy source and also in medicine, agriculture, heavy industry and in dating ancient materials.

Review questions

1. **(a)** What do you understand by 'radioactive emission'?
   (b) Describe the nature and properties of three types of radiation emitted by naturally occurring radioactive materials.
   (c) Describe three applications of radioactivity.
   (d) State briefly what you understand by 'nuclear fission' and 'nuclear fusion'.

2. **(a)** Outline briefly three methods which may be used to detect nuclear radiations.
   (b) Draw a labelled diagram of a Geiger–Muller detector.
   (c) Discuss the hazards created by radioactive materials and describe the precautions that should be taken in their use.

3. **(a)** What do 238 and 92 represent in the symbol $^{238}_{92}U$? How many neutrons, protons and electrons does a neutral atom of uranium-238 contain?
   (b) Balance the following nuclear reactions by finding the value of x in each case:

   (i) $^{210}_{84}Po \rightarrow {}^{x}_{82}Pb + {}^{4}_{2}He$
   (ii) $^{4}_{2}He + {}^{9}_{4}Be \rightarrow {}^{12}_{6}C + {}^{1}_{x}n$
   (iii) $^{4}_{2}He + {}^{14}_{x}N \rightarrow {}^{17}_{8}O + {}^{1}_{1}H$
   (iv) $^{1}_{1}H + {}^{7}_{3}Li \rightarrow {}^{4}_{2}He + {}^{x}_{2}He$
   (v) $^{27}_{13}Al + {}^{4}_{2}He \rightarrow {}^{x}_{15}P + {}^{1}_{0}n$
   (vi) $^{230}_{93}Np \rightarrow {}^{239}_{94}Pu + {}^{0}_{x}e$
   (vii) $^{232}_{90}Th \rightarrow {}^{228}_{x}Ra + {}^{4}_{2}He$

4. A radioactive material has a half-life of 20 days.
   (a) What do you understand by 'half-life'?
   (b) How long will it take for three-quarters of the atoms originally present to disintegrate?
   (c) How long will it take until only just one-eighth of the atoms originally present remain unchanged?
   (d) If an isotope of radium has a half-life of 1600 years, what fraction of a sample of this isotope is left after 9600 years?

11 Electronics

Learning objectives

After completing the work in this chapter you will be able to:
1. state the differences between conductors and insulators
2. define intrinsic and extrinsic semi-conductors
3. explain doping in semiconductors
4. explain the working of a p–n junction diode
5. sketch current–voltage characteristics for a diode
6. explain the application of diodes in rectification.

Electronics, which became a significant branch of technology less than 100 years ago, is concerned with electrical circuits designed:
- to process, store and transmit information;
- to sense and respond to changes in the surroundings;
- to control machines.

The development of electronics has made possible modern appliances such as television sets, radios, burglar alarms, digital thermometers, hi-fi systems, CD players, digital watches and computers. Modern electronic devices are based on scientists' understanding of electricity and the properties of conducting materials. This has allowed scientists to develop electronic components, such as diodes and transistors, to manipulate electric currents. The changing currents, or **signals**, in a circuit carry information to represent the strength of a sound wave, for example, or to trigger an alarm in response to a rise in temperature. This chapter introduces the properties of the materials used to make electronic components, in particular those of **semiconductors**, and examines the characteristics and applications of the semiconductor diode.

11.1 Conductors, semiconductors, insulators

Materials used to construct electronic components may be classified as conductors, semiconductors or insulators. The differing electrical properties of these materials depend on the strength of the force that holds the outermost electrons to the atoms from which the materials are composed.

Conductors

Conductors are materials with low electrical resistance. The outermost electrons of the atoms in a conductor are so loosely held that they become detached and move freely through the material. They are called **free electrons**. Metals such as copper and silver are examples of good conductors, but all metals conduct electricity.

The resistance that metals do have is caused by collisions between the moving free electrons and the vibrating atoms. At higher temperatures, the atoms in a material vibrate more vigorously and collisions become more frequent. This is why the resistance of a conductor increases with increasing temperature.

Insulators

Insulators are materials with very high electrical resistance. The outermost electrons are tightly held to their atoms – insulators therefore do not have free electrons. Consequently, they do not conduct electricity. Materials such as rubber, plastics and ceramics are insulators.

Semiconductors

Materials that are neither good conductors nor good insulators are called semiconductors. In a pure semiconducting material, such as silicon or germanium, the four outermost electrons of the atoms are more tightly bound than the outer electrons in a good conductor, but less tightly bound than those in an insulator.

At room temperature the random atomic vibrations associated with heat energy give a small fraction of these electrons sufficient energy to escape as free electrons, which can carry an electric current. An equal contribution to the current is made by the movement of positively charged **holes** through the material – see Fig. 11.1.

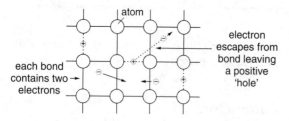

Fig. 11.1 Electrons and holes in a semiconductor

These holes are the bonds between atoms where an electron should be but is not, because it has escaped. Holes can hop from atom to atom – see Figure 11.2 – and respond to an electrical voltage in a similar way to electrons, but with the opposite charge.

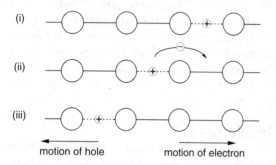

Fig. 11.2 How a hole moves

As the temperature of a pure semiconductor is raised the number of free electrons and holes increases. This means that the resistance of the semiconductor decreases with increasing temperature. If the temperature is lowered the numbers of free electrons and holes decrease, and the semiconductor's resistance increases again.

The band theory of conduction

In a single atom each electron has a specified energy – it is said to exist in an **energy level**. In a solid a material, where many atoms are close together, the energy levels merge together into **bands** of energy (see Fig 11.3). The bands have gaps between them – the gaps representing energies which electrons cannot have. The highest band is the conduction band – electrons with energies in this band can move freely through the material.

We can think of a metal as an array of positive ions immersed in a sea of electrons. The outermost electrons of the atoms occupy the conduction band and are not bonded exclusively to any one atom. The slightest potential difference across the metal will make the electrons flow. Metals are therefore good conductors and obey Ohm's law.

In a semiconductor, such as silicon, there is an energy gap between the filled levels and the conduction band. An electron in a bond between two atoms must receive extra energy to be lifted into the conduction band. The size of the gap is such that a significant number of electrons receive enough energy from random thermal (heat) vibrations to be excited into the conduction band. Raising the temperature promotes more electrons into the conduction band, so the resistance will fall with increasing temperature. Because semiconduction is a

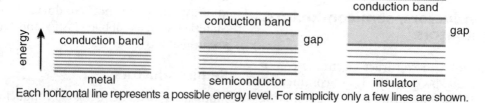

Each horizontal line represents a possible energy level. For simplicity only a few lines are shown.

Fig. 11.3 Conduction bands in a metal, a semiconductor and an insulator

property of the substance itself, materials such as pure silicon are called **intrinsic semiconductors**.

In some substances the gap below the conduction band is large and normal temperatures are not sufficient to excite electrons into it. There will never be any electrons in the conduction band – they all remain bonded to individual atoms and cannot move as a current. Such substances are called insulators. Table 11.1 compares carbon (in the form of diamond), silicon, germanium and tin, the first four elements of Group IV of the Periodic Table which provide a good contrast

Table 11.1 Band gaps and conduction in Group IV elements

Element	Gap size (eV)	Type
Carbon (diamond)	~5	Insulator
Silicon	1.12	Semiconductor
Germanium	0.72	Semiconductor
Tin	0.08	Conductor (metal)

Example

From the information in Table 11.1, predict which is the better conductor – silicon or germanium.

Answer

Germanium has the smaller band gap so electrons can cross it more readily. Germanium is the better conductor.

11.2 Intrinsic and extrinsic semiconductors

A pure crystal of silicon or germanium is an intrinsic semiconductor. Its semiconducting properties are determined by the band gap and the temperature. At all temperatures there are equal numbers of conduction electrons and holes to carry current.

To make practical electronic devices, the electrical properties of pure semiconductors are modified by adding small quantities of other elements, such as phosphorus and boron. This process is called **doping**. The modified semiconductors are called **extrinsic semiconductors** because their properties are determined by the doping elements that have been added from 'outside'. Fig. 11.4 shows how the addition of different doping elements may add extra conduction electrons or extra holes to carry current in the semiconductor.

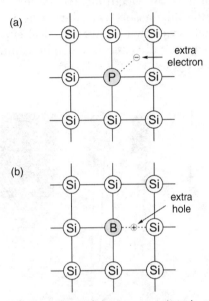

Fig. 11.4 The effect of doping a semiconductor
(a) Doping with phosphorous – n-type
(b) Doping with boron – p-type

Phosphorus has five outer electrons, four of which become involved in bonds to silicon atoms, leaving one extra electron to increase the electrical conductivity – as shown in Fig. 11.4(a). Because this extra electron carries a negative charge, a semiconductor doped with phosphorus, or similar elements, is described as **n-type**.

Boron has three outermost electrons. When silicon is doped with boron, these three electrons form bonds to neighbouring atoms, but a positive hole is produced by the absence of an electron to make a fourth bond – as shown in Fig. 11.4(b). Because the extra holes produced by boron doping carry a positive charge, a semiconductor doped with boron, or similar elements, is described as **p-type.**

The number of electrons or holes contributed by the doping atoms in semiconductors used to manufacture electronic components greatly exceeds the numbers in a pure semiconductor. Most of the current in an n-type material is carried by electrons; most of the current in a p-type material is carried by holes.

11.3 The p–n junction diode

A **diode** is an electronic component that acts rather like a one-way gate – it is a two-terminal device that allows electrical current to flow in one direction only. Some common diodes are shown in Fig. 11.5(a).

When pieces of p-type and n-type semiconductor are bonded together, or formed as layers on the surface of a semiconductor crystal, a p–n junction diode is produced. Fig. 11.5(b) shows the symbol for a solid-state diode.

If a potential difference is applied across the junction as shown in Fig. 11.6, current flows. The diode is then said to be **forward-biased**.

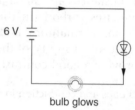

bulb glows

Fig. 11.6 (7.13) Forward-biased diode

When the diode is forward-biased, the p-type material is connected to the positive terminal of the voltage source. If the potential difference is applied in the reverse direction, virtually no current flows. The diode is then said to be reverse-biased (Fig. 11.7).

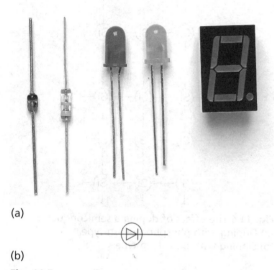

(a)

(b)

Fig. 11.5
(a) Diodes
(b) Symbol for a diode

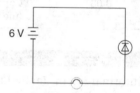

Fig. 11.7 Reverse-biased diode

Activity 11.1 *Experiment to investigate the characteristics of a diode*

You will need:
- a 2 V accumulator
- a rheostat
- a 10 Ω fixed resistor
- voltmeter
- milliammeter
- silicon diode

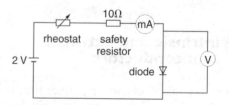

Fig. 11.8

Method

1. Set up the circuit as shown in Fig. 11.8. The safety resistor protects the diode by restricting the maximum current that can flow.

2. Adjust the rheostat to vary the voltage across the diode in steps of 0.1 V. Record the voltage and the corresponding value of the current through the diode.

(continued)

Activity 11.1 (continued)

3. Plot a graph of current against voltage.
4. Reverse the connections to the diode and repeat the current and voltage measurements.
5. Plot a second graph of current against voltage.

Discussion
The graph demonstrates that the diode does not obey Ohm's law – the current and voltage are not directly proportional. A much larger current flows when the diode is forward-biased than when it is reverse-biased.

The behaviour of a junction diode is explained by the formation of a very narrow layer, called the **depletion zone**, in the region where the p-type and n-type materials make contact – see Fig. 11.9. In this zone, holes from the p-type material combine with electrons from the n-type material and their charges cancel. The edge of the zone adjacent to the p-type material is left with a negative charge (because positive holes have 'disappeared'), and the edge adjacent to the n-type material becomes positively charged (because negative electrons have 'disappeared'). The negative charge repels the electrons in the n-type material, and the positive charge repels the holes in the p-type material. The junction thus acts as a barrier that electrons and holes do not cross.

The effects are shown in Fig. 11.10.

The effect of the bias on a diode can be shown by drawing a voltage–current graph. Such a graph is shown for a silicon diode in Fig. 11.11. The reverse current is only about one microampere (1 µA).

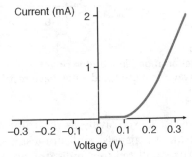

Fig. 11.11 Voltage–current curve for a silicon diode

Light-emitting diodes

Some diodes give out light when they are forward-biased and current flows. They are called light-emitting diodes (LEDs) – Fig. 11.12 shows the symbol for an LED. Like other diodes, an LED is usually protected by resistors connected in series to ensure that the current never becomes too large.

Fig. 11.12 Symbol for a light-emitting diode

An LED consumes much less power than a conventional torch bulb with a wire filament.

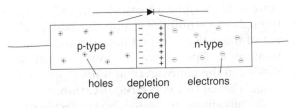

Fig. 11.9 The depletion zone in a p-n junction

When the diode is forward-biased the emf of the battery overcomes this barrier, and current flows. When the diode is reverse-biased the battery emf adds to the barrier increasing its 'height', and no current flows.

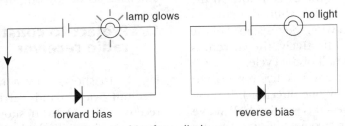

Fig. 11.10 Effects of forward bias and reverse bias for a diode

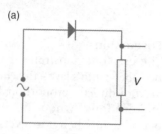

Fig. 11.13
(a) The diode as a rectifier
(b) The waveform produced – half wave rectification

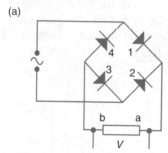
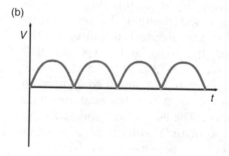

Fig. 11.14
(a) Diodes arranged in a bridge circuit
(b) The waveform produced – full wave rectification

LEDs are used extensively as indicator lamps on instrument panels and domestic appliances. The small red 'on' indicator on a television set is usually an LED.

11.4 Applications of diodes: rectification

The diode as a rectifier

An a.c. current flows first in one direction, then in the other. Such a current is no use if we wish to charge a battery, since its effect over time is zero. To overcome this problem, we can use the fact that a diode will pass current only in one direction. Look at the circuit shown in Fig. 11.13(a).

This shows an a.c. source, a diode and a load resistor. When the current is flowing clockwise, the resistance of the diode is low. In the second half of the cycle, when the current would flow anticlockwise, the resistance of the diode is very high. Effectively, the current is switched off in that half of the cycle.

Although the current has been **rectified** (made to flow in only one direction), it flows for only half the time – this is called **half wave rectification**. To overcome this we can use four diodes arranged in a bridge circuit as shown in Fig. 11.14(a). When the current is flowing clockwise it can only flow through diode number 1, and into the load at 'a'. It leaves the load at 'b' and flows back to the source through diode number 3. When the current flows anticlockwise it can only flow through diode number 2, into the load at 'a', out of the load at 'b', and back through diode number 4. Thus, there is always a current flowing in the same direction through the load – this is called **full wave rectification**.

The disadvantage of the bridge rectifier is that the current fluctuates from zero to its peak value many times per second – as shown in Fig. 11.14(b). To get round this a charge storage device, a capacitor, is connected in parallel with the load – the capacitor smooths the output from the rectifier. This is shown in Fig. 11.15(a) and its effect on the output in Fig. 11.15(b).

11.5 Project to construct a simple radio receiver

Fig. 11.16 shows how a radio signal in the medium wave band carries a signal to a receiver. Fig. 11.16(a) shows the basic radio wave, which is called the **carrier wave**.

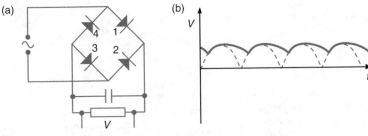

Fig. 11.15 The effect of a smoothing capacitor
(a) The circuit
(b) The waveform produced

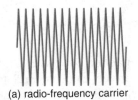 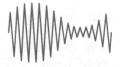

(a) radio-frequency carrier (b) modulating audio signal (c) carrier modulated with audio signal" AM

Fig. 11.16 AM radio transmission
(a) Carrier wave
(b) Sound signal
(c) Carrier wave with its amplitude modulated by the sound signal

For the medium wave band this is in the frequency range 300–3000 kHz – this is the **carrier frequency**. Different radio stations are transmitted on different carrier frequencies so that they can be separated by the receiver. Fig. 11.16(b) shows the **sound signal** produced by people speaking into microphones or by music from CDs. The amplitude of the carrier wave is modulated (changed) to follow the pattern of the sound signal as shown in Fig. 11.16(c). The overall effect is to produce an **amplitude modulated** (AM) signal.

The job of a radio receiver is to:
- pick up (receive) radio signals travelling through the atmosphere;
- separate the carrier signal frequency from all the other frequencies that are being received;
- separate the sound signal from the carrier signal.

Fig. 11.17 shows how a simple receiver performs these tasks in stages.

Radio signals are received by a metal antenna. The radio waves make the electrons in the antenna vibrate in the same pattern as the signal.

A particular carrier frequency is separated from all the others by a *tuned circuit*. This

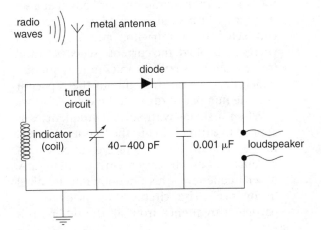

Fig. 11.17 AM radio reception

consists of a coil of wire and a capacitor connected in parallel. The coil is an inductor – an inductor is a wire coil with a core of air, iron or ferrite (an iron compound). The inductor carries a steady current in just the same way as a straight wire. However, it opposes any change in the current that flows through it. This results from the magnetic field created by the current. When the current changes, the field changes too and the change in the magnetic field induces an emf in the coil. The coil is said to have 'self-inductance'.

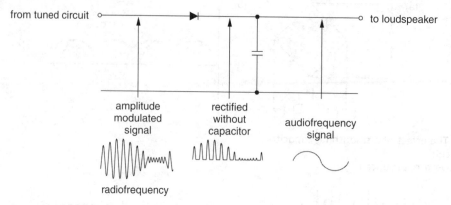

Fig. 11.18 Detecting the signal

By Lenz's law, the induced emf opposes the current change that brings it about.

The property of an inductor to oppose changes to the current it carries can be compared to the inertia of a moving mass which resists any change to its velocity. If a mass is suspended from a spring and released, the combination vibrates up and down at a set frequency. In a similar way, if an inductor is connected to a capacitor and a pulse of current is applied, the current oscillates to and fro around the circuit at a definite frequency, which depends on the value of the capacitor and the number of turns in the coil.

When a signal with several frequencies is applied to a tuned circuit, the frequency equal to the natural frequency of the tuned circuit produces a much larger current than any other frequency – this phenomenon is called 'resonance'. The circuit thus picks out a particular frequency from all the others. If a variable capacitor is used, the resonant frequency can be changed and the receiver tuned to pick up different radio stations.

Fig. 11.18 shows the signal from the tuned circuit – it is the carrier wave to which the circuit is tuned, modulated by the sound signal. If this signal is fed directly into a loudspeaker, nothing happens because the signal is symmetrical about the zero level. Any positive signal is balanced by an equal negative signal.

A diode and a capacitor are now used to separate the sound signal from the carrier wave. The diode rectifies the signal to remove half of the selected waveform. The capacitor acts as a frequency-sensitive resistor, having a much lower resistance to high frequency signals. The energy of the high frequency carrier wave thus passes to earth, leaving just the sound signal which can now be amplified and used to drive an earpiece or loudspeaker.

Activity 11.2 Constructing a simple radio receiver

You will need:
- copper wire
- a ferrite rod
- a variable capacitor
- a diode
- a capacitor
- a wooden board
- nails
- a crystal earpiece
- a long length of insulated wire.

Method
Fig. 11.19 shows a circuit design for a simple radio receiver which can be constructed with the components listed. Construct and test your own radio receiver using this design.

(continued)

Activity 11.2 (continued)

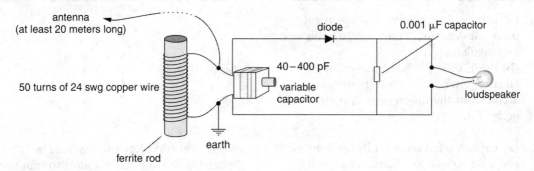

Fig. 11.19 A basic radio receiver

To make a working receiver you should:
- use as long an antenna as possible – stretch at least 20 m of wire in a horizontal line from a pole;
- establish a good earth connection – connect your earthing wire to a metal water pipe that runs underground, or to a heavy metal stake hammered deeply into the soil;
- make sure you scrape the varnish from the ends of the copper wire used to wind the coil;
- if possible, solder all the joints to ensure good electrical connections;
- do not overheat the diode when you are soldering it – you may damage it.

Summary

- Conductors are materials with low electrical resistance.
- Insulators are materials with very high electrical resistance.
- Materials that are neither good conductors nor good insulators are called semiconductors.
- In an intrinsic (pure) semiconductor, current is carried by electrons and positive holes – there are equal numbers of holes and electrons.
- The properties of semiconductors may be changed by doping with other elements.
- Extrinsic (doped) semiconductors are n-type or p-type – in an n-type material there are more electrons than holes; in a p-type material there are more holes than electrons.
- When pieces of p-type and n-type semiconductor are bonded together, or formed as layers on the surface of a semiconductor crystal, a p–n junction diode is produced.
- A diode acts as a one-way electronic gate – when the diode is forward-biased a current flows; when it is reverse-biased no current flows.
- Diodes can be used to rectify a.c. and to detect radio signals.

Review questions

1. **(a)** Distinguish between the electrical properties of conductors, semiconductors and insulators.
   (b) Use the simple band theory of electron energies in these materials to account for the differences in their behaviour.

2. **(a)** Explain what is meant by an 'intrinsic' and an 'extrinsic' semiconductor.
   (b) Explain the term 'doping' as used in production of semiconductors.
   (c) Describe how an electric current is carried in a p-type semiconductor and in an n-type semiconductor.
   (d) The resistance of a piece of semiconductor material is found to decrease significantly with a small rise in temperature. Is the material an intrinsic or extrinsic semiconductor? Explain your answer.

3. The figure which follows shows an incomplete circuit diagram for full wave rectification of an a.c. current.

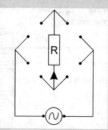

Copy and complete the diagram by inserting diode ciruit symbols so that the rectified current passes through R in the direction shown by the arrow.

4. **(a)** With the aid of a circuit diagram, show the difference between forward-biased and reverse-biased p–n junctions.
   (b) You are provided with a p–n junction diode, a variable resistor, an ammeter, a voltmeter and a d.c. supply.
   (i) Draw a circuit that could be used to determine the current–voltage characteristics of the diode.
   (ii) Sketch a graph to show how the current through the p–n junction varies with the p.d. across it.

Appendices

Appendix 1: Safety in the laboratory

In the laboratory:
- carry, don't throw;
- walk, don't run.

Remember:
- blades and broken materials can cause cuts;
- electricity can kill – especially in the wet;
- heat can cause burns;
- heavy things can bruise or crush;
- litter causes accidents, for example;
  - bags and clothing on benches or in gangways
  - pencils and ball bearings on the floor
  - untidily stacked furniture or equipment
  - paper waste near flames.

These are all avoidable problems.

Do:
- only what your teacher has suggested or agreed;
- only what you understand – if you are not sure, ask.

Afterwards:
- tidy your workspace.

If, in spite of everything, an accident happens then tell your teacher at once.

Appendix 2: Collecting and displaying experimental data

Fig. A1.1 Hazard warning symbols

Doing experiments involves making observations, recording results and writing up the experiments along with what they tell you. The more clearly you record your work, the more use it will be.
- If possible, record your work in an exercise book, not on single sheets of paper which are easily lost. If single sheets must be used, find some way of keeping them safely together.
- Start each record with the date and the title of the experiment.
- Write down exactly what you do, as soon as possible after you have done it. Follow the method that your teacher suggests. This may be to copy out the instructions step by step from a book or from the blackboard; or to stick a duplicated sheet of instructions into your book; or to write out a description of the method in your own words. If you do anything differently, or something unexpected happens, make a special note of it.
- Whenever you can, include a diagram of the apparatus.
- You often have to collect numerical data. These are best recorded in a table under headed columns. Before you start, decide how many columns you will need and

draw the blank table neatly. Remember to include columns for the quantities you will have to calculate. Never scribble observations – enter the results neatly in the right place as you collect them.

- When you do a calculation which has several steps, write out the steps clearly. This makes the calculation clear in your own mind. Also, if you make a mistake it is much easier to find it in a clearly explained calculation than in a scribble.

Example: a page from a laboratory notebook

To investigate the stretching of a spring: 11 November 2005

Apparatus used

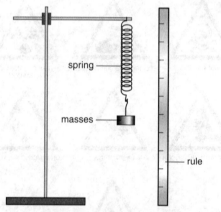

Fig. A2.1 Apparatus used to investigate stretching of a spring

Results

Table 1

Mass added (g)	Scale reading (mm)	Extension (mm)
0	596	0
10	601	5
20	606	10
30	610	14
40	615	19
50	621	25
60	625	29
70	630	34
80	635	39
90	638	42
100	640	44
110	641	45

Displaying results

Table 1 lays out the information collected from the experiment, but what conclusion(s) could you draw? Tables of numbers are not easy to interpret – graphs are much easier. Here is the graph of the results recorded in Table 1.

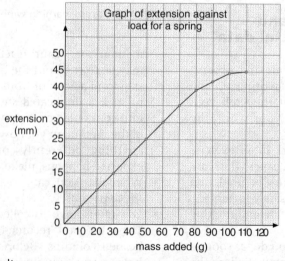

Fig. A2.2 Graph of results

Even in the first part of the graph, not all the results lie on straight line exactly – perhaps the experimenter could not always decide which scale division to take as the reading, perhaps the mass hanger was bouncing a little. It does not matter, the form of the graph is clear and the experimenter has drawn the best straight line through the points which seem to lie on a straight line, and sketched the curve which the later points seem to fit.

Note the following points about drawing graphs:
- every graph must have a title;
- draw both axes with a ruler;
- put the quantity controlled by the experimenter (the *independent* variable) on the *x*-axis;
- put the measured result (called the *dependent* variable) on the *y*-axis;
- choose scales for the axes which use up the whole of the graph sheet;
- label both axes with the quantity and its unit;
- the points are plotted so that they show up easily. A small point made with a sharp pencil does not show up well, and a large blob is not accurate; it is best to use a small vertical cross, +, or a circled point to mark the points;
- plot the points in pencil – you cannot rub out an ink mark if you make a mistake;
- if the graph is clearly a straight line, draw a best-fit straight line;
- if the graph is not a straight line, sketch a smooth curve which seems to fit the points – don't draw dot-to-dot;
- try to balance the points above and below your best-fit line or curve;
- remember that the origin, the point (0, 0), is not always a point on the graph – if it is there, it is just another experimental point; you must not force your best-fit line to go through it.

Interpreting results

A graph, like that shown in Fig. A2.2, can be used to answer a number of questions which might otherwise have been difficult to answer.

1. How is the extension related to the load?

Answer
The extension is proportional to the load up to a limit of 80 g. After that, the extension becomes less and less for the same change in load.

2. In the straight line region of the graph, what is shown by the slope of the line?

Answer
The slope gives the extension per unit load, or the stiffness. In this case a load of 70 g gave an extension of 34 mm so a load of 1 g would give an extension of $\frac{34}{70} = 0.49$ mm

3. What extension of this spring would you expect from a load of 35 g?

Answer
From the graph, 17 mm.
This is called *interpolation* – reading the graph between the known data points.

4. What extension of this spring would you expect from a load of 150 g?

Answer
You cannot tell. The graph shows that beyond 80 g the spring behaves in a different way, for which the experiment gives limited data. You can say that for a load of 120 g the expected extension might be about 46 mm, but more than this is dangerous! For all you can tell, the graph might start coming down again! Extending the line of a graph beyond the known data points is called *extrapolation*.

Appendix 3: Answering examination questions

It is not enough just to know the information for an exam. If you are to get the most marks, you also need to know how to use the information to answer the questions set by the examiner. Your can lose marks in an examination by not reading and understanding the question, and by not planning your answer.

Exam questions often start with cue words like 'Define ...' or 'State ...'. These words tell you what sort of answer is wanted. Here are some of these cue words.

- *Calculate or Find*

  'Calculate' means work out the answer to a numerical problem. Show all your working so that even if you make a mistake you will still get some marks. Make sure that you work in SI units. Make sure that you give the unit for the final answer.

  'Find' can be used instead of 'calculate' – in a practical, 'find' may mean 'measure'.

- *Compare*

  'Compare' means point out differences and similarities. Explain which are which. Don't just make two lists!

- *Define*

  'Define' means give a formal statement of the meaning of a term.

- *Describe*

  'Describe' means state in words, with diagrams where appropriate, the main points of the topic, in the right order. Look at the mark allocation to see how many steps are needed. Four marks, for example, suggests four steps in the description.

- *Describe and explain*

  'Describe and explain' means give a reason for each point you give in your description.

- *Draw a graph*

  'Draw a graph' means plot the given data on graph paper. Look at *Appendix 2* to see how to do this.

- *Draw the apparatus*

  Sketch it as well as you can and label it fully – don't waste time doing beautiful pieces of artwork.

- *Explain*

  'Explain' means 'give a reason for'. It does NOT mean 'describe' and you will not get marks for a description. You are being asked to answer the question 'Why does X happen?'

- *Give three reasons/examples*

  This is telling you how many reasons (or examples), which are *different* from each other, to give. There are no extra marks for giving more. Make sure that each one is different.

- *Name*

  'Name' means exactly that. Give the name of the object, process or quantity and nothing else.

- *Outline*

  'Outline' means give a brief overview of the topic. The number of marks and the space allowed on the answer paper will help you to decide how much is needed. Restrict your answer to the essentials.

- *Predict*

  'Predict' means that you are not expected to know the answer from prior knowledge; you get it by applying logic from information given in the question.

- *Deduce*

  'Deduce' means that you need to use some information given, or something you have worked out, to arrive at some decision or other.

- *State*

  'State' means give a brief answer but don't add any supporting arguments or reasons.

- *Suggest*

  'Suggest' either means that there is no right answer and you have to make a choice, or that you have to apply your knowledge to a situation you have not studied before. Try to think of a similar situation which you have studied and work from that.

Appendix 4: SI units

Base units

The International System (SI) of units has seven base units, and a variety of other units derived from these seven base units. The base units are listed in Table 2.

In order to indicate multiples of a unit, we use prefixes – the common ones are listed in Table 3. For example, a metre is the SI unit for length, but we can prefix this with 'kilo' to make kilometre, which is used to measure large lengths; or we can prefix it with 'milli' to make millimetre, for use with short lengths.

Table 2 SI base units

Quantity	Unit name	Abbreviation
Mass	kilogram	kg
Length	metre	m
Time	second	s
Electric current	ampere	A
Temperature	kelvin	K
Luminous intensity	candela	cd
Amount of substance	mole	mol

Table 3 SI prefixes

Prefix	Abbreviation	Meaning	Value
pico	p	one million-millionth	0.000 000 000 001
nano	n	one thousand-millionth	0.000 000 001
micro	µ	one millionth	0.000 001
milli	m	one thousandth	0.001
centi	c	one hundredth	0.01
deci	d	one tenth	0.1
kilo	k	one thousand	1000
mega	M	one million	1 000 000
giga	G	one thousand million	1 000 000 000
tera	T	one million million	1 000 000 000 000

Greek alphabet

Many derived units, and some of the SI prefixes, use letters from the Greek alphabet as abbreviations. The complete Greek alphabet is given in Table 4.

Table 4 The Greek alphabet

Lower case	Upper case	Name	Pronounced
α	A	alpha	
β	B	beta	'beater'
γ	Γ	gamma	
δ	Δ	delta	
ε	E	epsilon	
ζ	Z	zeta	'zeeta'
η	H	eta	'eeta'
θ	Θ	theta	'theeta'
ι	I	iota	'eye-ohta'
κ	K	kappa	
λ	Λ	lambda	'lamda'
μ	M	mu	'mew'
ν	N	nu	'new'
ξ	Ξ	xi	'ksy'
o	O	omicron	'oh-micron'
π	Π	pi	'pie'
ρ	P	rho	'roe'
σ	Σ	sigma	
τ	T	tau	'tore'
υ	Y	upsilon	
φ	Φ	phi	'fie'
χ	X	chi	'ky (as in sky)'
ψ	Ψ	psi	'p-sigh'
ω	Ω	omega	

Appendix 5: Significant figures

The accuracy of a calculated result can never be higher than that of the least accurate measurement within it. The original data and the answers are always given to the same number of significant figures.

Table 5 Examples of significant figures

Two significant figures	Three significant figures	Four significant figures
1800	1780	1783
0.0084	0.008 43	0.008 429
510	508	508.6
0.68	0.678	0.6780
96 000	96 200	96 180

Index

Page numbers in **bold** indicate tables. Page numbers in *italics* indicate figures.

aberration
    chromatic 19
    spherical 2
a.c. generators 57, 58, *58*
acceleration, centripetal 25
accommodation 13
aerials 46
alpha decay 116, *116*, *117*
alpha particles 109
alpha (α) radiation 108–9, *108*, **110**
alphabet rule 57, *57*
alternating current 57
    frequency 57, 66
alternating current (a.c.)
    generators 57, 58, *58*
alternators *see* a.c. generators
amplitude modulated (AM)
    radio
    transmission 129, *129*
angular displacement 23, *23*
angular velocity 23–4
anodes, in CRO 80
antennae 46, 49, 129
aperture 16, *16*
aqueous humour 12
Archimedes' principle 35–7, *37*, 38
    applications 39–40
armature winding 58
astronomical telescope 18–19, *18*
atom, structure 104–5, *104*
atomic bomb 113
atomic number 105, 116, 117

background radiation 117–18
balloons 39
band theory 124–5
bands
    conduction 124–5, *124*
    energy 124–5 *124*
    gaps in 124–5, **125**
banking (race tracks) 31, *31*
Becquerel, Henri 104
beta decay 117, *117*
beta particles 109
beta (β) radiation 108, *108*, 109, **110**

bicycle dynamo 59
binocular vision 13
blind spot 12
bolometers 46
boron, in semiconductors 125
bridge circuit of diodes 128, *128*
    use of smoothing capacitor 128, *129*
burglar alarm 101

cameras 15–16
    night vision 46
cancer treatment 47, 118–19
capacitor
    in bridge circuit 128, *129*
    in radio receiver 130
carbon-14 119
carrier frequency 129
carrier wave 129
cathode 78
cathode ray oscilloscope (CRO) 80–6, *81*
    design 80
    as measuring instrument 80–1
    sensitivity 82
    time base signal 82, 83, *83*
    trigger control 83
    uses 82–5
        frequency comparison 84, *84*, *85*
        phase comparison 83–4, *84*
        voltage measurement 82–3, *82*
        waveform display 83
        X input 83
        Y input 82–3
        Y-shift control 82
cathode ray tube 78, *78*, *80*
cathode rays
    production 78–9
    properties 78–80, *78*, *79*
centrifuge 31–2, *31*
centripetal acceleration 25
centripetal force 24–9
    factors affecting 26–8
chain reaction 112–14, *113*

chromatic aberration 19
ciliary muscles 12–13
circuit breakers 71, *71*
circuit symbol, transformer *60*
circular motion, uniform 22–33
    horizontal 29
    on tracks 30–1
    vertical 29
climate change 49
cloud chamber 111, *111*
commutator 58
compound microscope 14–15, *14*
conduction bands 124–5, *124*
conductors
    electrical 123
    see also bands, gaps
control rods 114
cookers 73
coolant 74
cornea 12
cost of electrical energy 75
CRO 80–6, *81*
    design 80
    as measuring instrument 80–1
    sensitivity 82
    time base signal 82, 83, *83*
    trigger control 83
    uses 82–5
        frequency comparison 84, *84*, *85*
        phase comparison 83–4, *84*
        voltage measurement 82–3, *82*
        waveform display 83
        X input 83
        Y input 82–3
        Y-shift control 82
crystallography, X-ray 47
Curie, Marie 104
Curie, Pierre 104, 108

data, experimental
    collection 133
    display 133–5
d.c. dynamos 58, *59*, *59*
d.c. generators 57, 58–9

see also d.c. dynamos
decay, exponential 107
decay constant 107
deflecting plates, in CRO 80
depth of field 16
detectors of radiation *see* radiation detectors
diffraction, X-rays 47, 90, *90*
diffusion cloud chamber 111, *111*
diodes 126–8, *126*
　bridge circuit 128, *128*
　　use of smoothing capacitor 128, *129*
　characteristics 126–7
　light-emitting (LEDs) 127–8
　　symbol *127*
　p-n junction 126–7
　　depletion zone 127, *127*
　　forward-biased 126, *126*, 127, *127*
　　reverse-biased 126, *126*, 127, *127*
　in radio receivers 130
　as rectifiers 128, *128*, *129*
　symbol *126*, *127*
direct current (d.c.)
　generators 57, 58–9
　see also d.c. dynamos
displacement, angular 23, *23*
domestic appliances 72–4
　power ratings **72**
domestic electrical safety 74–5
domestic wiring system 69–72, *73*
　circuit breakers 71, *71*
　circuits 71–2
　fuses 71, *71*, 72
　plug wiring 69–70, *70*
　switches 70–1, *71*
doping of semiconductors 125, *125*
double insulation 70
　symbol *70*
dynamos 58, 59, *59*

earth wire (plug) 70
earthing 70
eddy currents 60
Einstein, Albert 94, 112
Einstein's equation 98
electric current, effects on human body 75
electric lamps 72–3
　fluorescent 73
　incandescent 73
electrical appliances 72–4

electrical conductors 123
electrical energy, cost 75
electrical insulators 123, 125
electrical power stations *see* power stations
electrical safety, domestic 74–5
electromagnetic induction 51–65
　applications 61–3
　laws 52–3
　mutual 56–7, *56*
　used in generators 57–9
electromagnetic radiation 44–50
　applications 47–9
　detection 45–7, **47**, 110–12
　sources 45
　see also electromagnetic waves; *individual types*
electromagnetic spectrum 44–5, *44*
electromagnetic waves 44–5
　properties 45
　see also electromagnetic radiation
electromotive force *see* emf
electron(s) 79, 105
　charge 80
　energy levels 124
　free
　　in conductors 123
　　in semiconductors 124, *124*, 127
　mass 80
electron beam, in CRO 80, 81
electron gun 80
electron volt 95–6
electronic appliances 74
electronics 123–31
　band theory 124–5
　signals 123
　see also diodes; radio reception; radio transmission; semiconductors
emf, induced 52
　a.c. generator 58
　d.c. generator 58–9
　moving wire 54, 55, *55*
energy bands 124–5, *124*
energy gaps 124–5, **125**
energy mass equivalence 112
equations, nuclear *see* nuclear equations
examination questions, answering 135–6
exponential decay 107
extrinsic semiconductors 125–6
eye 12–13
　accommodation 13

binocular vision 13
defects 13
far point 13
near point 13
persistence of vision 13
eye lens 12
eyepiece
　of microscope 15
　of Newtonian telescope 19
　of refracting telescope 18

f-number 16
far point 13
Faraday, Michael 51
Faraday's law 53, 55
Fermi, Enrico 115
field strength, magnetic *see* magnetic flux density
filament, in CRO 80
film, photographic 15, 46
film badges 90, *90*, 110
fission, nuclear 112–15
　chain reaction 112–14, *113*
　energy production 113–15
　induced 112
　moderator 113, 114
　plutonium 113
　uranium-235 112, 113, 114
Fleming's left-hand rule 55
Fleming's right-hand rule 55, *55*
floating and sinking 37, 39, *39*
flotation, law of (principle of) 37–8
fluorescence 46, 47–8, 79
　caused by radiation 105, 112
fluorescent lamps 73
fluorescent light tubes 46, *46*
flux, magnetic 53, *53*
flux density, magnetic 54
flux leakage 60
focal length 2
　determination 9–12
　　by estimation 9
　　by lens formula 9–11
　　by lens-mirror method 11–12
focusing ring 16
food
　preservation 119
　radioactive materials in 118
force, centripetal 24–9
　factors affecting 26–8
free electrons
　in conductors 123
　in semiconductors 124, *124*, 127

frequency comparison 84, *84*, *85*
friction, maximum (limiting) 30–1
fuses 71, *71*, 72
fusion, nuclear 115–16, *115*

gamma camera 118
gamma emission 117
gamma (γ) radiation 108, *108*, 109, **110**
gamma rays *44*, 45, 47
gamma sterilisation 119
Geiger-Muller (GM) counter 47, 111–12, *111*
  to detect background radiation 118
Geiger-Muller (GM) tube 111–12
generators 57–9
  a.c. (alternating current) 57, 58, *58*
  d.c. (direct current) 57, 58–9
  see also d.c. dynamos
germanium 124, 125
global warming 49
graphs
  drawing 135
  interpolation 135
Greek alphabet 138, **138**
greenhouse effect 48–9, *48*
grid, in CRO 80
grid system 69

Hahn, Otto 113
half-life 106–8, **108**
hazard warning symbols *133*
heaters 73
hydroelectric power plants 66
hydrogen bomb 115
hydrometer 39–40, *40*, 41–2, *42*
hypermetropia 13, *13*
hysteresis loss 60

ignition coil 62, *62*
image formation
  camera 15, *15*
  compound microscope 14, *14*
  lenses 3–4
  magnifying glass 14, *14*
image properties 4–7, **6**, **7**
incandescent lamps 73
induced emf 52
  a.c. generator 58
  d.c. generator 58–9
  moving wire 54, 55, *55*

induction, electromagnetic see electromagnetic induction
induction coil 61–2, *61*
inductor 129–30
infrared radiation (IR) *44*, 45, 46, 48
insulators
  electrical 123, 125
  see also bands, gaps in
interpolation 135
intrinsic semiconductors 125
ionisation by radioactive substances 105
ionisation detectors 47, 110–12
  cloud chamber 111, *111*
  Geiger-Muller (GM) tube 111–12
  scintillation counter 47, 112
  spark counter 110–11, *111*
iris 12
iron core 60
isotopes 105
  of carbon 104, 105

kilowatt-hour (kW h) 75

lamps
  electric 72–3
  fluorescent 73
  incandescent 73
lens formula 9, 16–17, *16*
lenses 1–21
  concave see lenses, diverging
  converging 1–6, *1*
  convex see lenses, converging
  diverging *1*, 2, 7
  focal length see focal length
  image formation 3–4
  image properties 4–7, **6**, **7**
  optical centre 2
  ray diagrams 2, 4–7
  thin, definition 2
  virtual foci 2
  wide-aperture 2
Lenz, Heinrich 53
Lenz's law 53, *54*, 57
light bulbs 72–3
light-dependent resistor (LDR) 99, *99*
light-emitting diodes (LEDs) 127–8
  symbol *127*
light-sensing circuit 99, *99*
light waves 44
lighting circuit 71–2, *72*
Lissajous' figures 84, *84*, *85*

live wire (plug) 69–70
long sight 13, *13*
luminous paint 105

magnetic field strength 54
magnetic flux 53, *53*
magnetic flux density 54
magnetic hysteresis loss 60
magnification 8, 14, 18
magnification formula 8, 16
magnifying glass 14
mass energy equivalence 112
mass number 105, 116, 117
medical tracers 118
medium waveband 128–9
microphone, moving-coil 62–3, *63*
microscopes
  compound 14–15, *14*
  simple 14
microwaves *44*, 46, 49
Millikan, Robert 80
moderator 113, 114
mutual induction 56–7, *56*
myopia 13, *13*

n-type semiconductors 125–6
natural radiation 117–18
near point 13
neutral wire (plug) 70
neutrons 105
  in fission 112–13
Newtonian telescope 19, *19*
Newton's first law of motion 24
Newton's second law of motion 22
night vision cameras 46
no parallax method 3, 11
nuclear accidents 115
nuclear energy 113–15
nuclear equations 116–17
  alpha decay 116, *116*
  balancing 116
  beta decay 117, *117*
  gamma emission 117
nuclear fission 112–15
  chain reaction 112–14, *113*
  energy production 113–15
  induced 112
  moderator 113, 114
  plutonium 113
  uranium-235 112, 113, 114
nuclear fusion 115–16, *115*
nuclear power stations 113–15
  pressurised water reactor 114–15, *114*

Index 141

nuclear reaction 105
nuclear reprocessing plants 114–15
nucleons 105
nucleus 104–5
nuclides 105

objective (object lens)
  of microscope 15
  of telescope 18–19
oil drop experiment (Millikan) 80
optic nerve 12
optical centre 2

p-n junction diode 126–7
  depletion zone 127, *127*
  forward-biased 126, *126*, 127, *127*
  reverse-biased 126, *126*, 127, *127*
p-type semiconductors 125–6
pacemakers 119
peak value *66*, 67
persistence of vision 13, 81–2
phase comparison 83–4, *84*
phosphorus, in semiconductors 125
photocell 93
photoconductive devices 99
photoelectric effect 93–102
  applications 98–100
  demonstration 93–4, *94*
  description 93
  Einstein's equation 98
  energy of electrons emitted 98
  explained by quantum theory 94–6
  factors affecting emission 96–7, *96*
  investigations 93–4, 96–7, *96*
  photoconductive devices 99
  photoemissive devices 98, *98*, *99*
  photovoltaic devices 100, *100*
  threshold frequency 95, *95*, 97
  work function 95, **101**
  see also photons
photoemissive devices 98, *98*, *99*
photographic film 15, 46
photographic plates 110
photomultiplier tubes 46, 112
photon model of light 95
photons 94, *94*

energy 97–8
photoresistors 46
photovoltaic cell 46
photovoltaic devices 100, *100*
Planck constant 95
Plimsoll, Samuel 39
Plimsoll line 39
plug wiring 69–70, *70*
plutonium 113
power circuit 72, *72*
power lines 67–8
  a.c. model using transformers 68
  d.c. low-voltage model 67
power loss in transmission 67
power ratings, domestic appliances **72**
power stations 66–9
  hydroelectric 66
  thermal 66
power transmission 67–9
pressure and upthrust 37
pressurised water reactor 114–15, *114*
primary coil 59–60
principal axis 2
principal focus 2
protons 105
pupil 12

quantum theory 94

race tracks, banking 31, *31*
radar 49
radian, definition 22, *22*
radiation
  background 117–18
  natural 117–18
radiation detectors 110–12
  cloud chamber 111, *111*
  Geiger-Muller (GM) tube 111–12
  ionisation detectors 47, 110–12
  photographic plates 110
  scintillation counter 47, 112
  spark counter 110–11, *111*
radio reception 129–31
  antennae 129
  detection of signal *130*
  receiver 129, *129*, 130–1, *131*
  tuned circuit 129–30
radio transmission
  amplitude modulated (AM) 129, *129*
  carrier frequency 129

carrier wave 129
  medium waveband 128–9
  sound signal 129
radio waves *44*, 45, 46, 49
radioactive decay 105
  decay constant 107
  rate 107, *107*
radioactive isotopes 104, 105
radioactive waste 115
radioactivity 104–21
  agricultural uses 119–20
  applications 118–19
  characteristics of radioactive substances 105
  detectors of radiation 110–12
  hazards 117–18
  industrial uses 119
  medical uses 47, 118–19
  properties of radiation 108, *108*, **110**
  safety 110, 118
  types of radiation 108–9, **110**
radio-carbon dating 119
radon gas 118
raster scan 81
ray diagrams 2, *2*
  camera 15
  compound microscope 14
  converging lens 4, *4*, 5
  diverging lens 7, *7*
  magnifying glass *14*
  rules for construction 4
  telescope
    astronomical *18*
    Newtonian *19*
    terrestrial *18*
real-is-positive sign convention 17
rectifiers 128, *128*, 129
reflecting telescopes 19
refracting telescopes 18–19
relative density 38–40
reprocessing plants 114–15
resonance in tuned circuit 130
retina 12
ring main 72, *72*
Röntgen, William 88
root mean square (rms) value *66*, 67
Rutherford, Ernest 108, 109

safety
  electrical 74–5
  in the laboratory 133
  with radioactivity 110, 118

scintillation counter 47, 112
secondary coil 60
selenium photocell 100, *100*
semiconductors 46, 123–6
    doping 125, *125*
    extrinsic 125–6
    free electrons in 124, *124*, 127
    holes in 124, *124*, 127
    intrinsic 125
    materials 124
    n-type 125–6
    p-type 125–6
    resistance 124
    see also band theory; bands
ships 39
short circuit 71
short sight 13, *13*
shutter 16
SI units 137
    base units 137, **137**
    prefixes **137**
sign convention 17
significant figures 138, **138**
silicon 124, 125
simple microscope 14
sinking and floating 37, 39, *39*
slip rings 58
sound signal 129
spark counter 110–11, *111*
spark plug 62
specific gravity *see* relative density
spectacle lenses 13
speed of electromagnetic radiation 45
speed of light 45
spherical aberration 2
spur system 72, *72*
step-down transformer 61
step-up transformer 61, *61*
sterilisation, gamma 119
stopping potential (voltage) 96
Strassmann, Fritz 113
Sun, fusion process 115
suspensory ligaments 12–13
switches 70–1, *71*

telescopes 18–19
    astronomical 18–19, *18*
    Newtonian 19, *19*
    reflecting 19
    refracting 18–19
    terrestrial *18*, 19
television scan *81*
television tube 80, 81–2
terrestrial telescope *18*, 19
tesla (T) 54–5
thermal power plants 66
thermionic emission 79, *79*
Thomson, J.J. 80, 109
threshold frequency 95, *95*, 97
time base signal 82, 83, *83*
tracers, medical 118
transformers 59–61
    circuit symbol *60*
    energy loss 60
    simple designs 63–4, *64*
    step-down 61
    step-up 61, *61*
    structure 59–60, *60*
    tappings (taps) 61, *61*
trigger control 83
tuned circuit 129–30
    resonance in 130
TV scan *81*
    see also raster scan
TV tube 80, 81–2

ultraviolet radiation (light) *44*, 46, 47–8, 73
unit (commercial) of electricity 75
upthrust
    definition 35
    on submerged objects 35–7, *36*
uranium-235 112, 113, 114

velocity, angular 23–4
virtual foci 2
visible light *44*, 45, 46, 48, 73
vitreous humour 12
voltage measurement, using CRO 82–3, *82*

Wall of death (motorcycle stunt) 25, 29–30, *30*
water and electricity 74
watt 75
wave equation 45
wavebands 49
    medium 128–9
waveform display, using CRO 83
weight loss, apparent 37, *37*
wiring system, domestic 69–72, 73
    circuit breakers 71, *71*
    circuits 71–2
    fuses 71, *71*, 72
    plug wiring 69–70, *70*
    switches 70–1, *71*
work function 95, **101**

X input 83
X-ray tube 88, *88*, 89
    energy changes 89
X-rays 44, 45, 46–7, 79, 88–92
    absorption 90
    dangers 90
    diffraction 47, 90, *90*
    effect on photographic film 88
    energy 89
    frequencies 88, 89
    hard 90
    intensity 89
    interference 90
    penetration 90
    production 88
    properties 90
    soft 90
    speed 90
    uses 46–7, 90–1
    wavelengths 88, 89

Y input 82–3
Y-shift control 82